AMATEUR
BOAT BUILDING

Printed in the United States of America

ISBN: 978-0-9820733-1-5

Digitally Reproduced in 2012 by
CONVERPAGE Digital Reproductions
23 Acorn Street
Scituate, MA 02066
www.converpage.com

AMATEUR BOAT BUILDING

MICHAEL VERNEY

Foreword by
MAURICE GRIFFITHS, G.M., A.I.N.A.
Editor, *Yachting Monthly*

First Edition . . . 1948

PREFACE

THE text of this book is a modification of that which appeared as a series of articles in *Yachting Monthly* during 1946. I must first express my indebtedness to Mr. Maurice Griffiths, the Editor of that magazine, for permission to re-publish these articles and for the use of the majority of the illustrations. To this I must add my sincerest thanks for the help and advice which he gave during the writing of the original script and for contributing a Foreword to this book.

I am glad to have the opportunity of thanking Mr. Stanley Ogilvie, of Stanog Seacraft, Ltd., who, whilst he cannot be held responsible for any shortcomings which may be evident in this book, very helpfully criticized some of the original articles. I must also thank Mr. A. Khan, of Ahmadi Shipyard, Bombay, for allowing me to publish some of his ideas and methods, which have been thoroughly proved under tropical conditions.

As many books on boat building have already been published, it is felt that some explanation should be given for a further work on the same subject. The majority of these books treat only one type of boat, or one kind of construction, and, as the prospective amateur builder should know something about every aspect, he is often unable to decide which class of boat is most suited to his requirements and ability until he has studied quite a number of volumes. The necessity for such an extensive study is liable to subdue a man's desire to start work on his first boat, so this book has been written to include the building details of practically all types of yachts, excluding dinghies (which follow by common sense from an understanding of the larger types), and

PREFACE

steel construction. Consequently, to keep the length of the book within reasonable limits, few complete drawings and building instructions of *specific* vessels have been included. Instead, the building processes of all the different kinds of craft have been described in such a manner that the amateur is able to apply them easily to the specification and drawings which he will purchase before commencing work.

The "*Yachting Monthly* 3-Tonner," a boat specially designed for amateur building, is fully described, and some of the drawings are reproduced to a much reduced scale, providing an example of typical naval architects' plans.

All those who have had the opportunity to make yachting their hobby agree that "messing about in boats" is the finest possible pastime, but few of these people realize that the enjoyment can be increased tremendously when the owner has built the boat with his own hands. Most amateur boat builders agree that fashioning their own craft has given them as much pleasure as the subsequent sailing of her—a statement proved by the fact that so many of these folk have not been content at building just one boat!

I hope that this book will, in all cases, accomplish its task—to instil the would-be boat builder with the necessary incentive to start the venture, and to provide him with the information he requires to interpret the naval architects' drawings and instructions. Furthermore, I hope that the casual reader, having no previous intention of attempting such work, will, after perusing these pages, realize how enthralling and yet, to one with a fair amount of practical ability, how simple it can be.

M. P. VERNEY.

GLOUCESTER,
December 1947.

FOREWORD

AMONGST the hundreds of books which have been published on Yachting it has been only too rare to meet one that is really comprehensive. This has been particularly so in books on boat building. In most volumes on this truly fascinating subject the authors have confined their notes to one form of construction only, in some cases to one type and size of boat. This may be sufficient for the reader who has decided to build just the boat as described and no more, but the building of small craft is a very wide subject and the amateur cannot learn too much about boat construction as a whole.

Even the different types of wooden construction accepted for normal boats these days are many and varied. In the hull alone, for instance, we have a choice of clinker, carvel, double diagonal, or single spline planking, with sawn frames, joggled frames, steamed and bent timbers, stringers or seam battens ; we can build our hull with a normal round bilge or with chines and a flat bottom like a barge, or with a dead-rise or **V**-bottom with one or more chines aside. Such is the wide choice of construction methods open to the builders of small boats.

It has long been the policy of the *Yachting Monthly* to publish articles that are as comprehensive and informative as they can be on the subject covered, and in the series of articles that Mr. Verney prepared specially for us and which now form the contents of this volume, all the most usual methods of construction are described in detail. The intending builder will find here much of the really practical side of the craft, the hints and tips and useful ways of doing things that are nearly always lacking from amateur-written books on the subject and are usually only to be picked up at first hand in a boat builder's shop.

FOREWORD

A book of this kind entails an immense amount of patient work, as any writer of a technical book will know. Mr. Verney is to be commended for the thoroughness with which he has tackled his subject and for the excellence of his numerous diagrams.

However comprehensive a text-book may be, there are always critics who have other methods which they consider better than the author's. Boat building lends itself to an infinite variety of methods and dodges, and all possible ways of doing everything could not be adequately described in any single volume.

Some readers may suggest, for example, that as timber is so difficult to obtain for building small craft, why not do as the Dutch yards have done since their own supply of home-grown oak failed many years ago, and turn to steel construction? Alternatively, we can obtain sheets and angle strips of non-corrosible light alloy, such as Birmabright, and avoid troubles with timber.

Building in steel or light alloy, however, is another technique altogether and not to be picked up from a text-book alone. The methods of cutting and shaping, riveting or electric welding require a knowledge of metal working and a practised hand in the tools employed. A chapter on metal construction would be entirely inadequate as well as out of place in a volume on wooden boat construction.

This book should become the standard text-book in this country for amateur boat building, and with its aid I hope to see many people eventually sailing in their own craft and enjoying the work of their own hands. Boat building is in the Briton's blood, it is one of the most ancient of our industries, while the sailing of small craft keeps fresh in mind our noble heritage—the Sea.

MAURICE GRIFFITHS.

CONTENTS

Part I—HARD CHINE CONSTRUCTION

CHAP.		PAGE
I.	GENERAL INTRODUCTION	1
	Other books. Types of construction. Designs and lines drawings. Lofting.	
II.	FIRST STAGES IN BUILDING	16
	The stem. The frames and transom. Setting-up.	
III.	ERECTING AND PLANKING	30
	The keel and chines. Propeller shaft boring. Planking.	
IV.	MISCELLANEOUS JOBS AND TOOLS . . .	47
	Tools. Clamps. Steaming. Uncaulked and double-diagonal planking. Finishing. Caulking. Bottom painting. Topsides planking. Turning over.	
V.	COMPLETING THE HARD CHINE RUNABOUT .	64
	Deck framing. Carlines. Engine bearers. Rudder mechanism. Motor installation. Final planking. Bulkheads. Decking. Engine hatch. Interior fittings.	

Part II—ROUND BILGE CONSTRUCTION

VI.	TIMBERS, TOOLS AND SOME USEFUL HINTS .	81
	Choice of timber. List of tools. Types of fastenings. Glues and stopping. Painting and varnishing.	
VII.	BUILDING A 3-TONNER—(1)	98
	Types of hull. Types of planking. Dinghies. Clinker planking. Building of 3-tonner. The ballast keel. Building site and erection. Making the moulds.	
VIII.	BUILDING A 3-TONNER—(2)	117
	Shelves and ribbands. Bending the frames. Planking. Alternative planking methods. Keel. Deck framing. Rudder. Centreboard. Mast and spars. Motor installation. Launching.	

CONTENTS

CHAP.		PAGE
IX.	CONSTRUCTION OF A LARGER HULL—(1) . . Composite construction. All-metal hulls. Sawn frames. Building a 48-ft. cutter. Labour arrangements. Building site. The mould loft floor. The timber order. Seasoning. Slash-sawn and rift-sawn timber. Start of building. Scarphing the main members. Keel bolts. Copper and zinc sheathing. Plank rabbet.	139
X.	CONSTRUCTION OF A LARGER HULL—(2) . . Scarphs. Keel bolts. Sheathing. Sawn frames. Cant frames. Trenails. Floors. Keel notches and keelson. Building equipment. Counter construction. Shelves and planking. Joints and scarphs in planks. Plank fastenings. Deck structure. Building with straight-grained timber.	154

PART III—MISCELLANEOUS

XI.	VENTILATION AND CABIN CARPENTRY	. . 179
XII.	SHEATHING, BALLAST, CAULKING AND STOPPING	184
XIII.	A HALF-MODEL	190
XIV.	ELECTRICAL INSTALLATION	198

APPENDIX

| I. | TABLES . . . . | 211 |

PHOTOGRAPHS

PLATE		PAGE
1.	Planking-up a hard chine (or V-bottom) hull	6
2.	Construction of an 18-ft. hard chine hull well under way	7
3.	The first two frames are fixed to the stem	26
4.	A chine gusset with the frame set-up	27
5.	The bottom planked and sheer battens fixed	43
6.	The seam battens are notched into each frame	43
7.	The sheer batten in position	59
8.	The hull nearly completed, with covering boards in place	65
9.	A trial fitting of the engine hatch	65
10.	The finished boat	79
11.	A keel, stem and sternpost assembly	104
12.	Ribbands and frames in place	121
13.	Garboard and sheer strakes fitted	123
14.	Planking a double-ender—one ribband still in place	127
15.	Sternpost and keel, showing scarphs	152
16.	Views of a " bread and butter " model during assembly	191

DRAWINGS

FIGURE		PAGE
1.	The lines essential for the builder and a table of offsets	10
2.	How to find the right bevel on the frames	14
3.	An 18-ft. Hard Chine Runabout	16
4.	The stem, scarph and fastenings	19
5.	Cutting the stem rabbet	21
6.	How the frames are built up	22
7.	The chine and chine batten	23
8.	How the hood ends of planking are notched into the transom	25
9.	The keel joints (shown keel uppermost)	31

DRAWINGS

FIGURE		PAGE
10.	Section of false keel	33
11.	The stem fastening	35
12.	Boring for the propeller shaft	37
13.	Carvel and double-diagonal planking	39
14.	Method of marking garboard planks for cutting	40
15.	Methods of cutting planks	45
16.	Planking at the chines	46
17.	Clamps and cramps	48
18.	How to tighten a plank seam	49
19.	The steaming box	50
20.	Positioning the beam ties	60
21.	The seam batten is notched into the sheer batten	61
22.	Make the cradles before turning over	62
23.	Deck beams and carlines	64
24.	A shelf is not necessary for our prototype	67
25.	The engine bearers	69
26.	Shaping the underside of bearer	69
27.	Alternative types of rudder	72
28.	The link allows the tension of cable to be maintained	75
29.	Methods of decking	77
30.	Tongued-and-grooved decking	78
31.	Honduras mahogany is difficult to plane	83
32.	A few " not so common " tools	88
33.	Nothing is so annoying as a badly ground screwdriver	89
34.	Coach screws, boat nails and bolts	90
35.	Five different types of keel construction	100
36.	Some construction details of the *Yachting Monthly* 3-tonner	106, 107
37.	Re-lofting a raked transom	109
38.	The stern assembly	110
39.	Temporary slipway	113
40.	Cradle for hull	113
41.	A typical mould	115
42.	Scaffolding frames	117
43.	Approximate ribband positions	119
44.	Plotting a smooth curve on plank	124
45.	Planking layout	126
46.	Spiling templates	129
47.	Clinker planking details	133

DRAWINGS

FIGURE		PAGE
48.	Lines of a 48-ft. cutter	143
49.	An example of scarphing the main members of a 48-ft. hull	147
50.	Locking adjacent surfaces	149
51.	Types of sawn frames	155
52.	Cant frames in the curve of the bow	157
53.	Five examples of oak floors	159
54.	Two types of metal floors	160
55.	Plate floors with keelson plate	161
56.	Construction of the counter	164
57.	A typical bending-form for the shelves, etc.	165
58.	Types of planking scarphs	168
59.	Deck and deck beam fastenings	171
60.	The 48-footer built with straight-grained timber	173
61.	Details of " crook-less " frames	174
62.	Straight-grained frames and floors	175
63.	Knees and double planking	177
64.	How to exclude water	180
65.	Ventilators	182
66.	The plank joints correspond with the waterlines	192
67.	Check the hull with templates	195
68.	Skeleton diagram of electrical wiring	200
69.	Charging panel—double battery	203
70.	Charging panel—single battery 'A'	204
71.	Charging panel—single battery 'B'	206

BIBLIOGRAPHY

Amateur Boat Building. By W. F. Crosbie. (U.S.A.). (Rudder Publishing Co. $3.00.)
Boat Building. By Howard I. Chapelle. (Putnam & Co. Ltd. 35s.)
Boat Building Materials and Methods. By A. H. Lindley-Jones. (P. Marshall & Co. Ltd. 5s.)
Boat Building Simplified. By H. J. Ashcroft. (Capt. O. M. Watts. 5s.)
Build a Boat. By John L. Hacker. (U.S.A.)
Building a Little Ship. By W. H. Johnson. (Allen & Unwin. 6s.)
Canoes, Dinghies and Sailing Punts. By H. J. Ashcroft. (Capt. O. M. Watts. 1s. 6d.)
Cruising Yachts : Design and Construction. By T. Harrison Butler. (Robert Ross. 5s.)
Lifeboats and Their Conversion. By C. E. Tyrell Lewis. (Witherby Ltd. 8s. 6d.)
Modern Boat Building. By Edwin Monk. (C. Scribner's Sons Ltd. 18s.)
Motor Boat Building. By Chapman. (W. H. Smith & Sons. 12s. 6d.)
Naval Conversions. By C. H. Evans. (Witherby Ltd. 8s. 6d.)
Practical Motor Boat Building. (British Motor Boat Mfg. Co. 12s. 6d.)
Practical Yacht Construction. By C. J. Watts. (Robert Ross. 21s.)
Rigging. By H. A. Calahan. (Macmillan & Co. Ltd. 12s. 6d.)
Simple Boat Building. By Geoffrey Prout. (Brown, Son & Ferguson. 5s.)
Small Boat Building. By Edwin Monk. (C. Scribner's Sons Ltd. 18s.)
Small Boat Building. By H. W. Patterson. (Capt. O. M. Watts. 5s.)
Small Sailing Craft Design and Construction. By J. F. Sutton. (Pitman & Sons Ltd. 7s. 6d.)
Wooden Boat and Ship Building. Van Gaasbeek. (U.S.A.) F. J. Drake & Co., Chicago.
Yacht Sails. By Terence North. (C. Scribner's Sons Ltd. 7s. 6d.)

PART ONE

HARD CHINE CONSTRUCTION

CHAPTER I
GENERAL INTRODUCTION

A LARGE proportion of yachtsmen, and especially small boat yachtsmen, are " practical " men ; that is, they take an interest in the construction and working of their yachts, they carry out all minor repairs, and can keep their vessels at sea under all normal circumstances. Every yachtsman schemes and plans his ideal boat, but only those owners who get her built to their own design ever seem to be satisfied. This is one of the reasons why yachting is held to be an expensive pastime, whereas in fact it can be made one of the cheapest. It is possible for the goal of possessing an ideal boat to be attained at very reasonable cost, and in a far more satisfactory and satisfying manner than by means of professional building, by undertaking this fascinating task oneself.

Amateur boat building has never been very popular in Britain, but this position will continue to change for as long as the demand for yachts exceeds the available supply, and prices remain abnormally high.

This pastime has been popularized in America, whereas in Britain the following reasons have hitherto retarded progress :

(1) The buying of small craft, up till 1939, was a fairly simple and cheap procedure.
(2) The trade is recognized as a very skilled one, and the amateur does not think himself capable of the task.

GENERAL INTRODUCTION

(3) So few books and articles have been published on the subject, that many potential enthusiasts have not realized that they were capable of attempting such work.

(4) Designs and lines drawings are very limited in type and are by no means plentiful. They are rarely seen in boating advertisements.

(5) Many enthusiasts have insufficient floor space available in which to build a boat.

(6) The proposed mooring spot may be far away and the foreseen transport difficulties, coupled with the aforementioned points, often make things appear rather impracticable.

Before continuing with the details of amateur building, the fact that it is well within the ability of any practical man must be proved. Providing the tyro has detailed building instructions and the correct design, there is no reason at all why he should not build a very strong and successful boat. The following are important points and the more consideration given to them, the better and quicker will construction be:

(a) Some help is valuable, but not essential. Too many hands should be avoided.

(b) The builder, and his helpers, must have great enthusiasm and determination, so that drawbacks do not cause despair.

(c) There is no limit to the number of tools which can be used, but the minimum will be listed later. Certain power tools, such as an electric drill and a small band-saw, although not essential, are great assets.

(d) Proficiency in handling normal carpenter's tools leads to good workmanship, and a general understanding of the principles of boat building

GENERAL INTRODUCTION

should be gained before the commencement of work.
(*e*) If no proper workshop is available, the boat can be built out of doors, but a shelter or roof of some sort is advantageous.
(*f*) A local shipwright can be very helpful and he will probably lend out certain rarely-used tools. Sometimes, however, his advice can be dangerous to the amateur, so that the latter should rely as far as practicable upon his own judgment and instructions.

To commence with, all the available books on the subject may be read. A concise bibliography of those publications normally available is given on p. xiv, and it will be apparent that nearly all these works specialize in one class of boat (perhaps hard chine only, or just bent frame construction) so that the prospective builder need only study those books which concentrate on the particular type of craft he has chosen.

By reading all or some of these books, the prospective boat builder will gain considerable confidence, as they all mention, in varying manners, the fact that the job is not so difficult as it appears. He will soon find his boat being built in his imagination, with all the various stages of construction arranging themselves in simple and orderly fashion. To convince the present reader still more, let it be said that every determined attempt at amateur boat building which the writer has known, has proved entirely successful, most being their builder's first endeavour.

One frequently reads accounts of such ventures in the yachting magazines, and several amateur-built vessels have been world-famous. For instance, Harry Pigeon, an American, built the *Islander* almost single-handed, and then sailed her round the world! Therefore, even a

GENERAL INTRODUCTION

deep-water sailing yacht, with her very strong framing, is within the reach of the amateur, but as we shall see later, although such a craft should be theoretically as simple to build as a dinghy, professional assistance is usually worth its cost in saving time, labour and materials.

In completing this section, let it be mentioned that like all things which a man fashions with his own hands, a boat so built produces never-ending satisfaction to her owner: he knows every detail of her and knows exactly what she can stand up to. It is only under these circumstances that the yachtsman can have full confidence in his boat; she is his " dream ship," and he probably acquired her for the cost of a couple of years' maintenance on " the other man's boat."

Types of Construction

Vessels for amateur boat building can be divided into two groups:

(1) Small craft, roughly under 30 ft. LOA.
(2) Big craft, over 30 ft. LOA.

A numerical division between these groups is difficult to decide, but there are many other considerations besides length. Tonnage is not a fitting rule, because a " small " sailing yacht might have a greater tonnage than a " big " motor-boat. One could differentiate by saying that a boat of group (1) does not possess grown timbers or floors, whereas group (2) is characteristic of vessels with a few or all frames and floors of grown crooks. The " small " group only will concern us for a while, so we will subdivide it into:

(*a*) Hard chine construction.
(*b*) Round bilge construction.

TYPES OF CONSTRUCTION

These types, in this order, will be treated later, from the amateur building point of view, but before going into detail there are a few items, normally the first ones to concern the prospective builder, which are common to all types of build, and will therefore be treated now.

Choosing the design is a foremost question. The tyro is not advised to contemplate too big a boat, unless reliable help is available, although as long as the required building site is at his disposal it is within the scope of his ability to build the largest craft of group (1). Inevitably, the battle between chine and round bilge construction crops up early in the procedure, so this position must be clarified.

Most amateur builders believe that a chine boat is the simpler to build. Certainly it appears so, but in practice, for most types of vessel, round bilge is found to be somewhat easier and definitely quicker, especially if the builder knows a little of the practical snags to be encountered. The truth of the matter is, that a hard chine hull is almost entirely straightforward carpentry, needing very little shipwright's skill to complete it, whereas a round bilge hull, with its curves, moulds and steaming, is far from conventional joinery.

If the tyro is not really confident in his ability, and time and expense are not controlling factors, then he is advised to build hard chine. Should he be unable to come to a decision even at this stage, the following should settle the matter.

Arguments about the relative merits of these hulls at sea are still popular, but the true position is this. For inland water motor cruising, hard chine build is difficult to excel, giving a more easily driven and faster hull. Round bilge scores decidedly at sea, with lighter weight, greater lateral resistance and no ledges for the seas to slap against. Round bilge proves more successful for all sailing yachts,

Plate 1.—Planking-up a hard chine (or V-bottom) hull.

Plate 2.—Construction of an 18-ft. hard chine hull well under way.

GENERAL INTRODUCTION

although on inland waters the smaller displacement hard chine sailers prove extremely successful. One further point must be made clear; the foregoing remarks concern the comfort of these boats at sea rather than their seaworthiness. A hard chine deep-water craft, if well designed, can ride out a storm with every confidence, but she will not do so with any degree of comfort. A chine-built sailing yacht has crossed the Atlantic Ocean, and the type is used far more in America than over here, even for ocean sailing.

As regards cost, the round bilge boat will always prove the cheaper, especially for a large hull.

Therefore, from all these considerations, it should be possible for readers to decide which type of build to adopt for the hull they have in mind. Both types are to be treated in this series of articles, so if any reader is still doubtful, the subsequent text will inform him of the exact constructional pros and cons in either case.

Designs and Lines Drawings

From personal requirements and a brief knowledge of the construction, a suitable set of drawings, together with a specification, must be procured. A very brief specification is sufficient, especially if the drawings are thorough, because details of scantlings and materials are variable and can be decided by the builder, who, having read all the instructions on boat building he can find, and paid a few visits to boat-yards, should have gained a reliable knowledge of " what will do " for each member.

The beginner is strongly advised not to prepare his own design. So many amateur-built craft, representing countless hours of thought and labour and scheming, have been disappointing when they have finally gone afloat owing to faults in the design that could have been rectified by anyone with the necessary experience.

DESIGNS AND LINES DRAWINGS

Before starting to build a boat, therefore, the amateur is wise to choose a suitable design prepared by a competent designer or naval architect, or to ask a professional designer to " vet " his plans for a small fee. It is easy to correct glaring faults in a design, but not so easy when the boat has been built.

Although many designs of yachts and dinghies and small sailing and power craft are published in the yachting magazines, it must be understood that all these plans are the copyright of the designers and a royalty fee must be paid and permission obtained from the designer of the chosen plans for any boat to be built to them. If the would-be builder " lifts " his plans from a published design without obtaining permission to build, he infringes the copyright law and becomes liable to the penalties attached to this form of irregularity. Some first-class plans of small craft, however, are offered by the various naval architects—the eight Yachting World designs, and the *Yachting Monthly* 16-ft. Sharpie and 3-Tonner are good examples for amateur construction—and there should be no difficulty in obtaining a suitable set of plans.

Fig. 1 shows a set of lines and table of offsets needed for building a small craft such as a dinghy or half-decked sailing boat. An experienced builder could manage with less, but the product, although a good boat, would probably be somewhat different in shape from that represented by the drawings. A Table of Offsets is very useful, and with the aid of these a boat can be built with only a very small sketchy set of drawings, because all the measurements needed are shown in this table and therefore do not have to be lifted from the drawings.

Conversely, if the drawings are on a big scale, between $\frac{1}{2}$ in. and 1 in. to 1 ft., then no table of offsets would be needed, the builder scaling off the measurements he

GENERAL INTRODUCTION

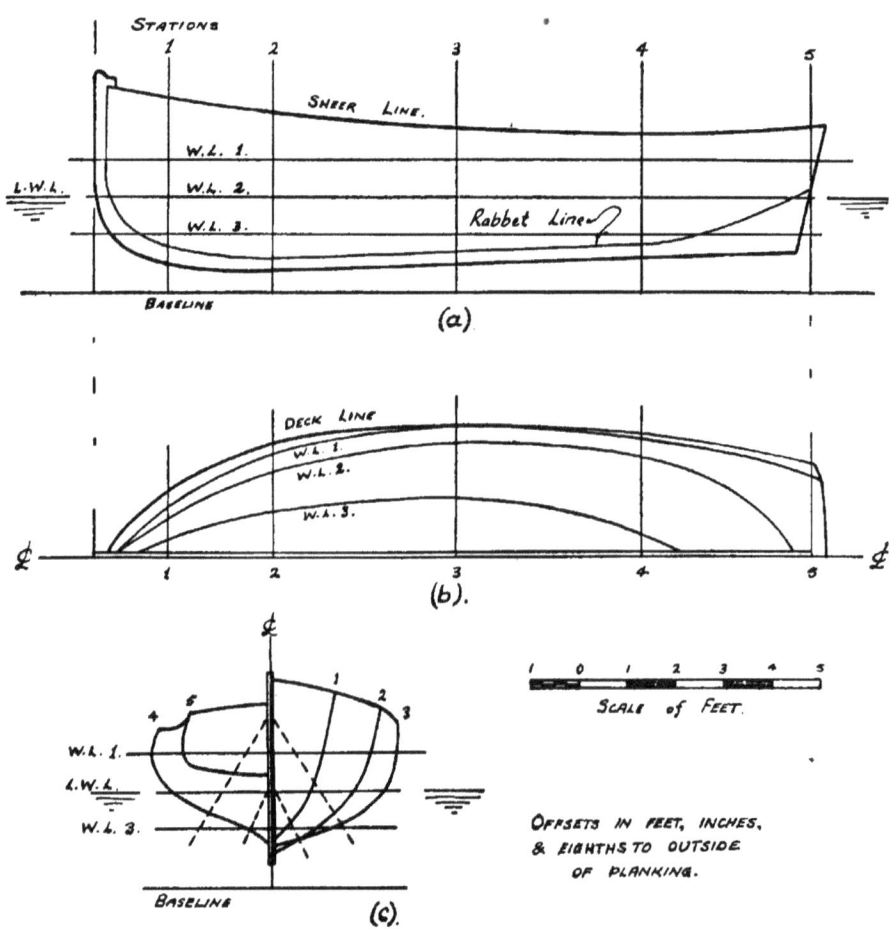

STATION	FROM BASELINE TO					FROM CENTRELINE TO			
	SHEER	W.L.1	W.L.2	W.L.3	KEEL	SHEER	W.L.1	W.L.2	W.L.3
1	3-11-1	2-8-1	1-11-2	1-2-1	0-11-4	1-3-7	1-0-6	0-9-5	0-3-3
2	3-7-3	2-8-1	1-11-2	1-2-1	0-11-3	2-3-0	2-0-6	1-8-1	0-10-6
3	3-3-7	2-8-1	1-11-2	1-2-1	0-9-7	2-7-3	2-7-3	2-3-6	1-2-4
4	3-2-3	2-8-1	1-11-2	1-2-1	0-11-6	2-4-6	2-6-1	2-0-7	0-6-4
5	3-4-3	2-8-1	1-11-2	1-2-1	—	1-8-4	1-10-1	—	—

(d)

FIG. 1.—The lines essential for the builder and a table of offsets.

DESIGNS AND LINES DRAWINGS

required during lofting. Offsets are nearly always given in feet, inches and eighths of an inch, to the outside of the planking.

The drawings of Fig. 1 are very brief, of course, merely intending to show the various lines required. For hard chine build, working drawings should have a station at each frame (or rib) and at least two waterlines above the chine. Round bilge requires a station at each mould to be used in building, and at least two waterlines above the load-waterline and two more below it. Hard chine construction needs fewer waterlines, because the bottom futtocks of the frames (the members between the chine and the keel) are normally straight, the topside futtocks only having curvature.

Drawing (*a*) is called the profile plan and shows the outline of the hull, with the baseline below it and the waterlines as horizontals, parallel to the baseline. The frame stations (numbers 1 to 5) are set at right angles to the baseline. It is normal to place them at equal intervals, but the frames themselves are not necessarily equidistant, because those which are for'ard of amidships are placed with their after faces on the station-line, those abaft amidships having their for'ard faces on the station-line. The reason for this is apparent from Fig. 2, as the bevel must be taken *off* the frames; it cannot be put *on*. The midships frame has no bevel and can be placed on either side of its station-line. Drawing (*b*) is the half-breadth plan, showing the hull as viewed from above, with the waterlines now as curves and the stations at right angles to the centreline. The body plan (*c*) represents all the frame stations as curves and the waterlines straight. The table of offsets (*d*) gives the distances of all points of intersection of lines from the centreline and from the baseline, so that when lofting (or drawing the lines full-size), the baseline and centreline can be run-in, all inter-

GENERAL INTRODUCTION

section points plotted, and these points joined by the various lines needed for building. A separate list of offsets is, of course, needed for the stern curve, but this is generally lifted straight from the profile plan.

Should the amateur decide to make his own drawings for a small motor-boat, there is no need for any calculations, and, providing he has a good eye for conventional proportions, there is little risk attached to this method. However, a better way would be to draw the design with reasonable accuracy and then get a naval architect to check it over and do a full set of drawings. This is essential for any type of sailing boat. An alternative way of preparing working drawings is to lift the lines from an actual hull, which is on the hard, or dry-docked. An interesting job but a tedious one, especially if done between tides!

When designing, start with the profile plan and get the desired outline sketched in above the baseline. Then put the proposed deck-line on the half-breadth plan and draw the required midship section with two intermediate sections on the body plan. From these sections, the waterline can be plotted roughly on the half-breadth plan, and without altering the midship section at all, any necessary adjustments must be made to the other sections, to produce fair waterline curves. For a small boat, the profile plan needs no correction, but for a bigger hull, a series of vertical lines across the body plan is represented by corresponding curved buttock lines on the profile plan, which naturally need adjustment each time a section or waterline is altered. When the remainder of the frame sections are drawn in (by transferring their offsets from the half-breadth waterlines to the body plan waterlines) they may not appear as fair curves, so the process of mutual adjustment between sections and waterlines must start all over again.

LOFTING

The job requires patience, but is intriguing and finally very satisfying.

Lofting

The object of lofting is to reproduce to full size all the lines needed for building, so that patterns or moulds can be lifted from them to the exact size. Lofting often presents difficulties to the amateur, mainly because he rarely has sufficient floor space to lay down the full set of lines, and they are best drawn on a floor. However, all difficulties can be overcome and for a small boat such as we are at present considering, there is no need to loft the full lines, but only the body plan, which takes up very little space. If a solid sawn keel has to be made, the profile plan will be necessary to get the shape of this, but even then only the outline of this member is required, so it can be laid down alongside the building site by plotting the offsets from baseline to keel. The stern can be laid down separately, as it is always scarphed to the keel. The half-breadth plan is really only needed for big boats with a shelf, so that this member can be steamed to shape before putting into place.

Nevertheless, the only way to draw the body plan accurately, especially for a round bilge hull, is to draw it in conjunction with the other plans, so that all should be lofted to get a perfect hull. The hard chine hull scores here, because, with the aid of about five offsets per frame (either tabulated by the designer or scaled off the drawing), a very accurate body plan can be drawn, and this is the only plan required for building one of these boats. More offsets are necessary for a round bilge hull and even then the full-size plotted points rarely join up into perfect curves, so that errors creep in, especially in the floors and bottoms of the frames, where they strike a very acute angle with the waterlines and intersections

GENERAL INTRODUCTION

occur less frequently. Good lines drawings incorporate diagonals in this region (shown dotted on Fig. 1 (c)), to enable accurate frame sections to be lofted, and the diagonals are represented similarly to waterlines, but on the opposite side of the half-breadth centreline. When lofting the body plan, it must be remembered that the line produced by joining up all the offset points represents the outside of the planking, so the actual frame line must be drawn afterwards, with a space equal to the plank thickness, plus the correction for bevel (shown in Fig. 2) which is obtained from the half-breadth waterlines.

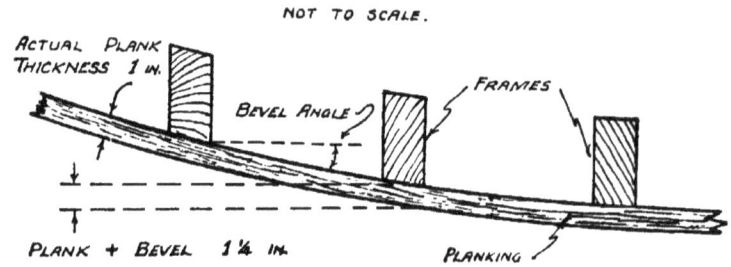

FIG. 2.—How to find the right bevel on the frames.

The best type of loft floor for small boats is a plain smooth timber floor, and the lines are best picked out with white paint to make them indelible. If no workshop is available and the boat is to be built out of doors, a level layer of concrete can be set on to firm soil and the lines painted on this in black. A tarpaulin cover will ensure preservation of the lines and the whole platform can easily be broken up after building. The workshop wall can be used, by fitting plywood or asbestos-cement sheeting against it and lofting the lines on to this, although, as will be shown later, the marking of patterns is simpler from lines laid down on the floor. A method incorporating the advantages of both floor and wall lofting, is to draw each frame on a separate sheet of strong paper,

LOFTING

which can be made up of several strips glued together, and pinned to the wall for plotting. A pattern can very easily be lifted from such a sheet.

A light wood spline, preferably of ash, is used for running-in the curved lines, and must be long enough to do the whole of one frame section in a single sweep.

CHAPTER II

FIRST STAGES IN BUILDING

TO commence actual building details we will take as a prototype a simple hull, an 18-ft. hard chine motor runabout, such as the one shown in Fig. 3. She has cockpits fore and aft, with the enclosed engine compartment separating them. Steering is from the for'ard

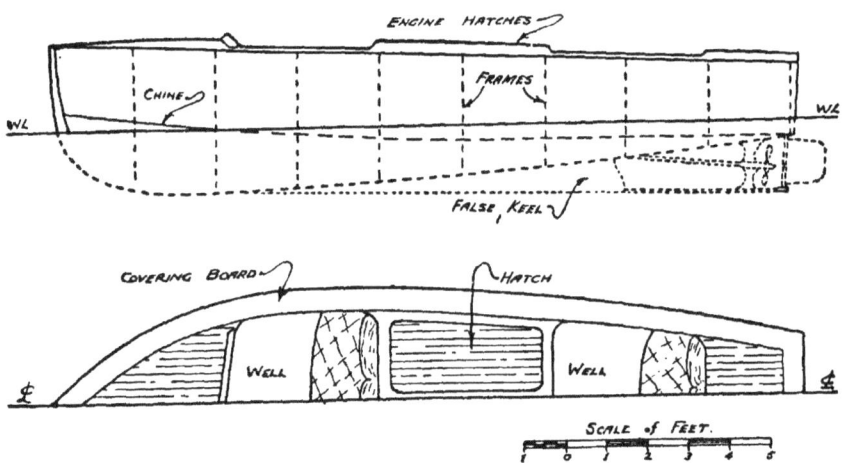

FIG. 3.—An 18-ft. Hard Chine Runabout.

well, and both wells have athwartships seats to accommodate two persons each with comfort, or three as a maximum. Her beam is 5 ft. 2 in., draught under the forefoot 1 ft. 6 in., and 5 in. at the transom.

Her frames are sawn from 1-in. oak and are spaced at 2-ft. centres. She is carvel planked, of ½-in. pine, but the different types of planking are to be discussed later, and several other methods could be used on our runabout. We will now work right through all the operations of building this boat, in their correct sequence, incorporating nearly every detail likely to be found in a hard chine hull. To start with, only the conventional construction will be dealt with, but optional methods of making each member will either be considered at the time, or, where they concern several parts, collectively at the end.

The Stem

Our prototype has no solid keel, but, as we shall see later, the keel is formed of two battens which are bent to the required curve; quite a conventional method with boats of this size and type and simplifying building considerably. When a solid keel is used, it is normal to fashion this first, and scarph the stem on to it; so keeping as near to this order as possible, we will make the stem first. This member really consists of two parts scarphed together, the upper piece called the " stem " and the lower one the " forefoot." It could be made of one solid piece of timber, but there are a number of objections to this practice. Firstly, as oak crooks have to be used to make any curved stem, difficulty arises in choosing a crook large enough to make the whole stem and keep accurately to the curve with its grain. Secondly, such long oak crooks, when as thin as $1\frac{1}{2}$ in. or 2 in., are very liable to split or " check " even a long time after cutting. Thirdly, in a similar manner, such a thin lamina will often warp or twist after shaping, which might easily cause leaks and weakening in so small a boat. Fourthly, the grain of oak is never consistent, so that even if a piece with the grain keeping to the right curvature is chosen, it will prove tough and knotty in parts, rendering difficult the cutting of the plank rabbet and of general shaping.

Oak crooks are cut from large limbs or curved trunks of the trees, and as these are often rejected by sawmills as " useless," one should visit a sawyer who specializes in boat-building woods, where the correct and well-seasoned pieces will probably be found. It will be apparent that

FIRST STAGES IN BUILDING

by making the stem of two sections, perfect crooks can be found, and the resultant member will remain sound and rigid. In boat building, the uniform width of a piece of timber is called its "sided" measurement, and the varying depth its "moulded" measurement. Our stem is sided 2 in. and moulded according to the construction profile plan. Fig. 4 (*a*) shows the complete stem-piece, together with its fastenings and adjoining members. The process of fashioning it should be as follows :

The outline of the stem-piece is represented on the loft-floor, and, if the inner edge is not shown it must be drawn in to get the moulded measurements. To make a pattern of it, nail several pieces of light planking together (about $\frac{3}{8}$ in. thick stuff) as sketched in Fig. 4 (*b*), so that one side is smooth and the other has the gussets on it. To mark the pattern, place a series of small nails or tacks (about $\frac{1}{2}$ in. long with big flat heads) all round the stem outline on the loft-floor, with their points upwards, and spaced at intervals of about 1 in. Then, with someone to hold each end of the pattern, lower it gently on to the nails (with its smooth side downwards) and tap it all round with a mallet. Lift away, and nearly all the nails will be found in the pattern, giving an accurate transfer of the outline. Mark all the nail holes with pencil, take out the nails, and join up all the marks with a pencil and spline. Cut out the pattern with a band-saw (frame-saw, or compass-saw) and finish to the lines with a spokeshave. Pencil on the line of the scarph. Now mark the upper piece on to its oak crook. Cut this out, leaving spare timber on each end, and then mark the scarph accurately across it.

The stem may be lofted on to a piece of strong paper, which can be cut out to the pattern, parted at the scarph, and tacked to each crook respectively. Great care must be exercised in cutting the scarph face, especially in

THE STEM

getting it square to the sides of the stem. The forefoot is shaped similarly, leaving some spare timber for the scarph, the position of which is marked from the finished half of the joint. In marking this, hold the pattern over the two crooks and orient them into their correct alignment. By covering one face of the scarph with chalk, and placing the other face to it, the " high-spots " can be determined, and chiselled down until a perfect fit is obtained. For this type of scarph a perfect fit is essential. If preferred, the wedged scarph shown in Fig. 4 (c) may

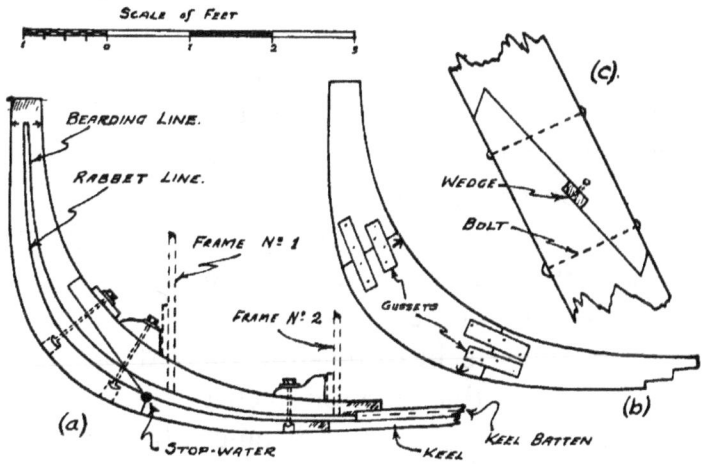

FIG. 4.—The stem, scarph and fastenings.

be used. The ends of this scarph interlock and only a smooth flat finish is required on the faces because the wedge (best made of teak, alternatively oak) pulls the two parts together, so that fastenings are almost superfluous, being used merely to stop sideways movement. The wedge has a very slight taper and is held in place by an oblique brass screw, indicated in Fig. 4 (c).

The best fastenings to use are galvanized steel carriage bolts, $\frac{3}{8}$ in. in diameter. Yellow-metal bolts are very variable in strength in the small sizes, but stainless steel ones are ideal and, if used, the diameter could be reduced

FIRST STAGES IN BUILDING

to $\frac{1}{4}$ in. The faces of scarphs should be luted with white lead paint or liquid marine glue to stop any water from entering the joint, and a dowel or " stop-water " of pine should be driven as shown in Fig. 4 (*a*) for the same purpose. The holes for the bolts can be drilled with a $\frac{3}{8}$-in. or $\frac{1}{4}$-in. auger, while the two parts are held rigidly together with clamps. An undersized hole is unsatisfactory, as the galvanizing of a bolt may be damaged during driving. Fastenings are much more easily bolted through the wedge type joint, as there is no tendency for the two halves to drift apart on tightening, as there is with the ordinary scarph. In the latter type, the bolts take the entire stress, both in sheer and tension, but it is still the conventional method.

After planing the sides of the whole assembly, the rabbet- and bearding-lines must be marked on. These lines are drawn on the construction plan of a proper set of plans, with offsets tabulated to simplify lofting. Alternatively, the rabbet-line must be drawn from the profile plan and the bearding-line plotted beside it, by drawing the plank thickness on each of the half-breadth waterlines and scaling off the distances between rabbet- and bearding-lines on the sides of the stem. These intervals can be transferred to the corresponding waterlines when marked on the stem. Next, the rabbet must be notched out, so that the hood ends of the planks can settle right into it. This notching is done by making a number of isolated slots at intervals of about 3 in. all along the rabbet, to the correct depths, to serve as guides for the notching of the remaining parts. The correct depth is indicated by means of a wood templet, the same thickness as the planking and about $\frac{1}{2}$ in. wide. When this templet sits in the notch to its full depth, touching both the rabbet- and bearding-lines, the notch is correct to the planking. The idea is shown in Fig. 5. This

THE FRAMES

figure also shows how the cutwater must be chamfered down to leave a $\frac{5}{8}$-in. wide face to the stem. Later, a

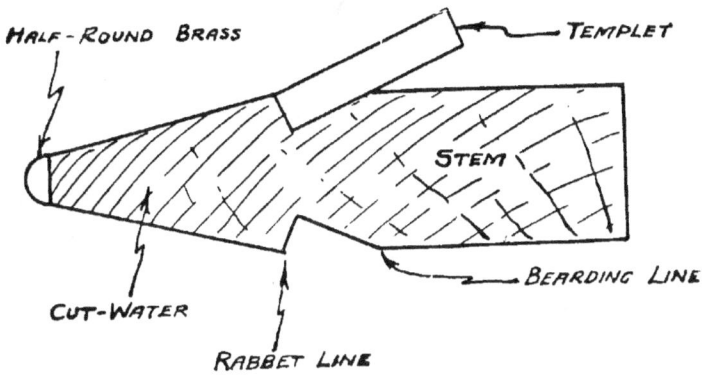

Fig. 5.—Cutting the stem rabbet.

strip of $\frac{5}{8}$-in. half-round brass is to be screwed along this face.

THE FRAMES

Our next task is the building of all the frames, which members correspond to the ribs of a small round bilge boat, and the timbers of a big craft. The frames of a hard chine boat are very strong, and can thus be placed at the large intervals of 2 ft. in our present example. They are sided 1 in. and moulded 2 in. above the chine, and increasing from 2 in. at the chine to 3 in. at the keel. Fig. 6 shows the midships frame. In all nine frames the topside futtocks have curvature, whereas the bottom futtocks are straight. Oak crooks are not required for these slight curves, although, if oak planking with curved grain can be obtained, the topside futtocks may with advantage be cut from it. Strength is supplied at the chine knuckles by $\frac{5}{8}$-in. oak gussets, fastened with three $1\frac{1}{2}$-in. No. 10 brass screws in each futtock, and luted with waterproof casein glue. The keel joint is similarly rein-

FIRST STAGES IN BUILDING

forced by ¾-in. oak floors, with four screws in each side. Natural oak crooks are again unnecessary for gussets and floors, but the straight grain must run parallel to the inboard side of these triangles.

Deck beams cannot be fitted until the hull has been partly planked, so temporary beam ties or cross spalls must be fitted to each frame to simplify setting-up. From Fig. 6 it will be seen that the underside of the beam tie is in line with the sheer mark (or top of the topside

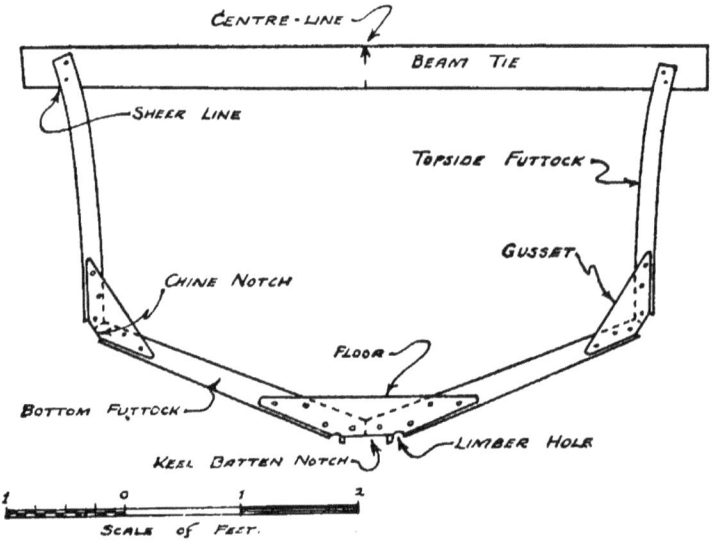

FIG. 6.—How the frames are built up.

planking) on both futtocks, and that it is securely screwed to the futtocks on the side opposite to the gussets and floors, enabling a plumb-line to be hung accurately from top to bottom of the frames. It should be 4 in. by 1 in. in scantling, although 2 in. by 2 in. would do. In the present instance, all the beam ties should be exactly the same depth, the reason for which will be given under "Setting-up."

A tricky point crops up in regard to the notches for the chine and keel battens. These must be cut during the building of the frames and therefore should be lofted on

THE FRAMES

the body plan or on the patterns. Now, the chine and keel battens are each of a constant width and central thickness throughout their lengths, so that although the widths of the frame notches will thus be constant, the depths of the notches will vary. The drawing in Fig. 7

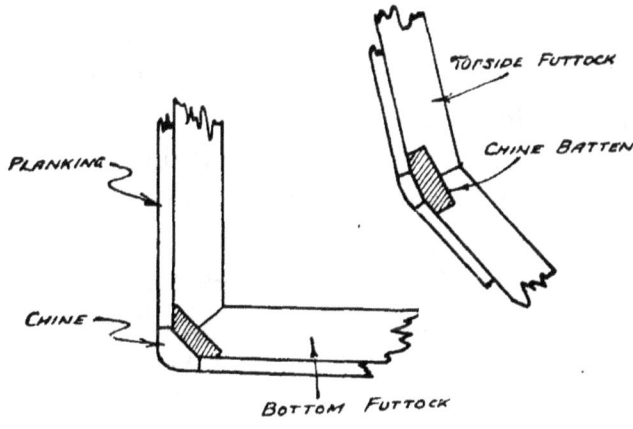

FIG. 7.—The chine and chine batten.

represents two chine joints (or knuckles) of very different angles, showing the reason for the above point. It may appear easier to cut these notches after the erection or setting-up of the frames, but all possible bench jobs should be done before erection when conditions are best for good workmanship. The method for obtaining the shape of the notches will be given in the next chapter, under the shaping of the chines and keel. The limber holes, to allow the free passage of water in the bilge, must be cut close to the keel batten.

As regards patterns for the futtocks, for each complete frame a single light wood mould of one half-frame is sufficient, so that the exact chine angle can be obtained when fastening the gussets. The exact angle at the floors need not be represented as a pattern, as the two finished half-frames can be placed together at their heels

FIRST STAGES IN BUILDING

and the distance between the two sheer marks at their tops checked to see that the angle at the floor is right. Having corrected this, the floors and beam ties can be screwed into place.

Only one snag attends the building of frames, and prevents them from being pure carpentry. This is the existence of bevels on all the outboard edges. These bevels arise from the fact that the planking takes a continuous curve along its length, whereas the frames are all parallel to each other and thus present different angles to the planking. Not only are these bevels all different, but the bevel on any one frame changes slightly from top to bottom of that frame. This bevel may be measured from the waterlines of the half-breadth plan for the topside futtocks, but there is normally so little bevel on the bottom futtocks, that it is either left until after erection, or put on by judgment. Bevel is best applied before cutting the chine and keel notches, as these notches follow the same bevel as the chine and garboard strakes of planking respectively. A carpenter's bevel (or adjustable square) may be used to test the angle, and a plane, spokeshave or band-saw (with canting table) for cutting off the required amount.

It will be recalled that the lines lofted on the body plan represent the frames at their widest, so that bevel must always be taken from those sides of the frames which are away from the station-lines. Where bevel is very slight, it should be left entirely until after erection. Then it will be applied with a plane or spokeshave during the "fairing-up" of all the other bevels. Amateur builders may prefer to put on all their bevels after setting-up. This method looks simple, but if good oak has been used it will prove very tiring work, in very awkward positions. It can be done with the after bottom futtocks, but the for'ard three or four have considerable bevel, and as this

THE FRAMES

angle is difficult to determine (unless the lines drawings incorporate diagonals), it should be estimated (slightly less than the topsides bevel) and applied roughly before erection.

The transom is a rather special frame, having its futtocks screwed to the transom planking, without gussets or floors. The futtocks still have bevel, and the transom planking overlaps the frame all round by an amount equal to half the thickness of the side planking. This means that the after hood ends of the side planking are notched

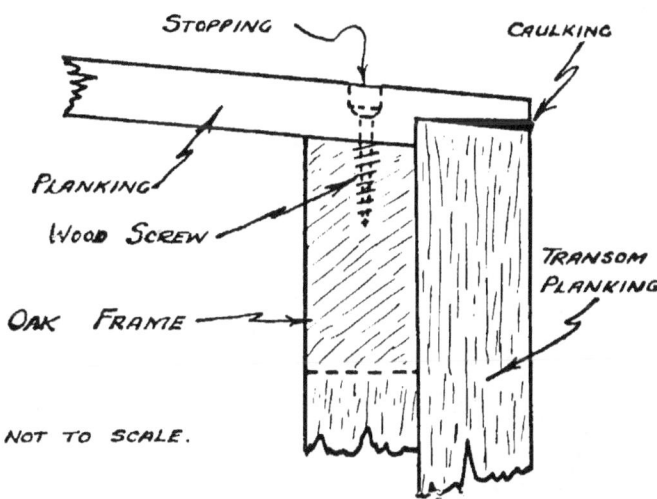

Fig. 8.—How the hood ends of planking are notched into the transom.

into the transom, as shown in Fig. 8, and screw-fastened to the oak frame only, leaving a caulked joint all round. The transom planking consists of a number of horizontal boards, under 8 in. in width, and at least 1 in. thick, with caulked seams between them. To stiffen the transom against bumps, and to prevent the boards from warping, a central vertical piece of 1 in. by 1 in. oak may be screwed to the inside of the planking. All brass screws are driven from the outside, through the boards to the oak framing, and paint or glue luting is used between all close surfaces.

FIRST STAGES IN BUILDING

The transom deck beam has a camber which gives a rise of about 1 in. in 20 in. span. Normally, the transom planking is finished flush with the top of the beam, but this depends upon the type of deck intended.

Before commencing to erect the framing, the first two frames must be secured to the stem; they notch over the stem to terminate at the bearding-line. The positions of these frames are shown dotted in Fig. 4 (*a*). No. 1 is

Plate 3.—The first two frames are fixed to the stem.

screwed to the chock which takes one of the stem scarph bolt nuts, and No. 2 to a special chock bolted to the after end of the forefoot. Fastenings should be four 2-in. No. 12 brass screws at Frame No. 1, and two similar screws through the floor of Frame No. 2. A check measurement from the stem head to the two beam ties ensures true normality to the baseline, but these two frames should only be temporarily fastened until all the frames have been assembled on the stringers, and checked —see later.

SETTING-UP THE FRAMES

The fact that our runabout example has a straight sheer will be apparent from Fig. 3. Certain fast motor-boats have a convex or reverse sheer, whereas most boats have the conventional curve giving lowest freeboard just abaft amidships. Only with the runabout type of hull does a straight sheer appear pleasing, and as erection is considerably simplified by making it straight, this sort of hull

Plate 4.—A chine gusset with the frame set-up.

is good for an amateur's first attempt or to serve as an example for building instructions.

The building site chosen, two 6-in. by 2-in. timber beams must be fixed above the floor, spaced about 3 ft. 6 in. apart, and extending just beyond the ends of the hull. These stringers must be perfectly straight, because the boat is to be erected on them with its bottom uppermost. By this means, work on the hull is facilitated and the structure may be supported at the most convenient height for working. The stringers should be supported on trestles, at each end and in the centre, to give complete

FIRST STAGES IN BUILDING

rigidity. Building is further simplified by giving the stringers a slope down from the after end to the for'ard end, so that the keel is brought more horizontal, leaving no high parts difficult to reach. In our present case, if the stringers are mounted 15 in. above a level floor at the transom, and 7 in. above at the stem, the waterline will be brought horizontal, so that all the frames can be accurately set-up with a plumb-line. Simple trestles, consisting of a cross-piece and four splayed legs, give good rigidity, and the legs can be skew-nailed to a timber workshop floor, or planted in holes or concrete if building out of doors. The stringers can be screwed to the trestles with the aid of cleats or angle brackets, but if building is to take place on a concrete floor, where anchoring of the trestles is impossible, the stringers must be fitted with diagonal braces between them, so that a firm structure is formed that can be moved about on the concrete floor without distorting.

The next job is to attach all the frames, the transom and the stem, to the two stringers. Firstly, mark the positions which the beam ties are to occupy, along the top of the stringers. The stem will need an athwartships batten attached to it, corresponding to the frame beam ties, and it can then be positioned on the stringers. The two frames should check dead vertical with a plumb-line, but an error less than $\frac{1}{4}$ in. is immaterial in this direction. The transom of our prototype will not be vertical, but as it is at right angles to the sheer-lines it can be set-up square to the stringers and cleated in that position.

A centreline string can now be stretched from each end of the boat, to touch the undersides of the beam ties, and the remainder of the frames can be positioned and cleated to the stringers. Finally, to hold the frames firmly in a vertical position, a wood batten must be nailed or screwed temporarily to each side of each frame, running the

SETTING-UP THE FRAMES

length of the boat and as close to the chines as possible. Starting from Frame No. 2, each frame must be plumbed in turn and fastened to the battens. A number of short battens would do equally well, as they can be removed completely as soon as the chine and keel battens are fixed.

CHAPTER III

ERECTING AND PLANKING

THE first two stages of building, namely, the making of the frame-work and the erection thereof, have now been treated. The next stage consists of adding to the inverted skeleton until it becomes a hull.

THE KEEL AND CHINES

These longitudinal members are each made up of two sections: the keel, consisting of the keel batten and the outer keel; the chines, consisting of the chine battens and the outer chines. The way in which these members notch into the frame knuckles was mentioned in the last chapter, so also was the fact that the shape of their sections changes continually along their lengths. On account of these changes, the sections must be lofted, or drawn full size, at each frame. Then the dimensions can be transferred to the actual timber, which should in all cases be shaped as accurately as possible on the bench; not on the boat. Lofting is best carried out on stiff paper, by transferring the angle of the knuckle from the body plan, or frame pattern, and plotting the shape of the batten on to it.

Taking the keel first, we find that both the width and the thickness of the keel batten remain constant throughout the length. Also, the outer keel behaves similarly, so that the surface where they touch each other cannot be of uniform width everywhere. Now, the floor futtocks at the transom actually form a straight line with each other, so both the keel batten and outer keel will be perfectly rectangular at this point. Therefore, by making

THE KEEL AND CHINES

a tracing of this joint, on thin typing paper, or proper tracing paper, it can be aligned over each of the paper frame knuckle drawings, and the outline of the keel section at each frame pricked through the tracing. On joining up these marks with pencil lines accurate loftings will be obtained. Fig. 9 (*a*) represents the keel joint at the transom, and Fig. 9 (*b*) shows clearly how the tracing

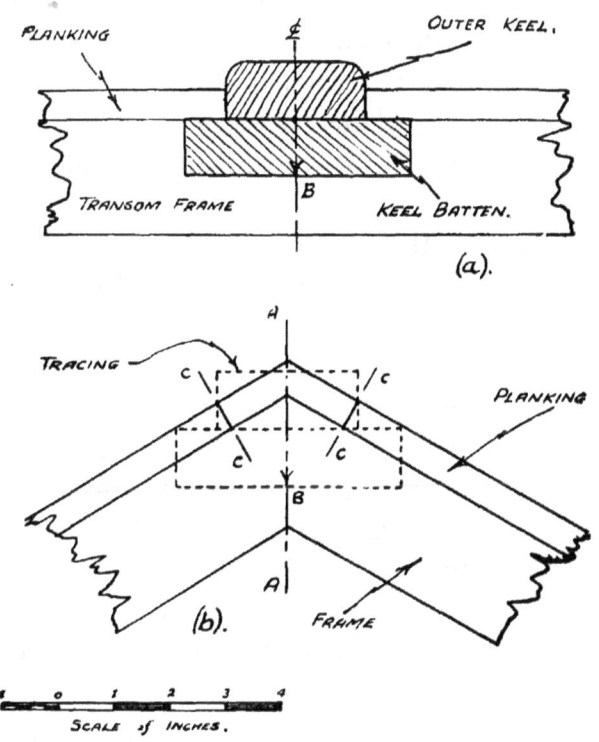

FIG. 9.—The keel joints (shown keel uppermost).

(of Fig. 9 (*a*)) is superimposed over a different frame. The planking is drawn on the other frames first, then the lines bisecting the keel joints (A—A) are put in. When the centreline of the keel batten (B) is on the line A—A of the joint concerned, the tracing is moved parallel to the line A—A until the edges of the planking (lines C—C) just touch the sides and base of the outer keel.

ERECTING AND PLANKING

This naturally requires some trial and error, as there is no mechanical way of determining just where the planking edges fit, but in practice it is very simple. On pricking through the essential points with a scriber, or other spike, the keel sections can be drawn in.

The dimensions of the two keel members are known, and as they are consistent in section the chosen pieces of timber can be sawn and planed to their exact sizes. Good, clean and well-seasoned English oak is the best easily-available timber to use. The keel batten must be fitted first, so having planed it to size, mark the centre of each frame across it. On these lines transfer the positions of the chamfers, which must be taken off (obtained from the lofting of the keel), and join up all these points to form four lines each running the length of the batten. Pencil lines will do, but a carpenter's spike will leave deep scratches which are far easier to work to. If a circular saw is available the chamfers may be skimmed off roughly with this before planing; otherwise, a plane alone must be used. Square off the end next to the stem and fit the batten along the frames. Small adjustments will be necessary, and these completed the member can be clamped into its final position, the transom end marked and cut, and the screw holes marked at each frame. On removal, the screw holes may be drilled and countersunk, and each frame notch well luted with paint or marine glue. The final fitting and screwing may then take place.

The outer keel is shaped in a similar fashion. Where the false keel is to be bolted on, the outer keel must be planed flat to take it, but for'ard and abaft this region the edges of the part protruding outside the planking should be rounded off. Screw intervals must be not less than 6 in. anywhere, except at the propeller shaft position and where the false keel fits. There is room for screws at each side of the shaft hole, but these should not be driven

THE KEEL AND CHINES

until the hole has been bored, to obviate any damage to the auger. Our keel batten is 4 in. wide by 1 in. thick, while the outer keel is 2½ in. wide by 1 in. thick, fastened with two 2-in. No. 12 brass screws at each frame through the batten (avoiding the floor to frame screws) and single 1¾-in. No. 12 screws holding the outer keel, at the above mentioned spacing. Although the false keel cannot be fitted permanently until the bottom has been planked, it should be made and tried in position next. Its profile can be taken from the boat drawings, and in section it appears as in Fig. 10, tapering slightly towards the shoe and bored to take slanting bolts. These bolts are to be ⅜-in. galvanized carriage bolts. Their borings are difficult to make, but a hand auger is quite satisfactory.

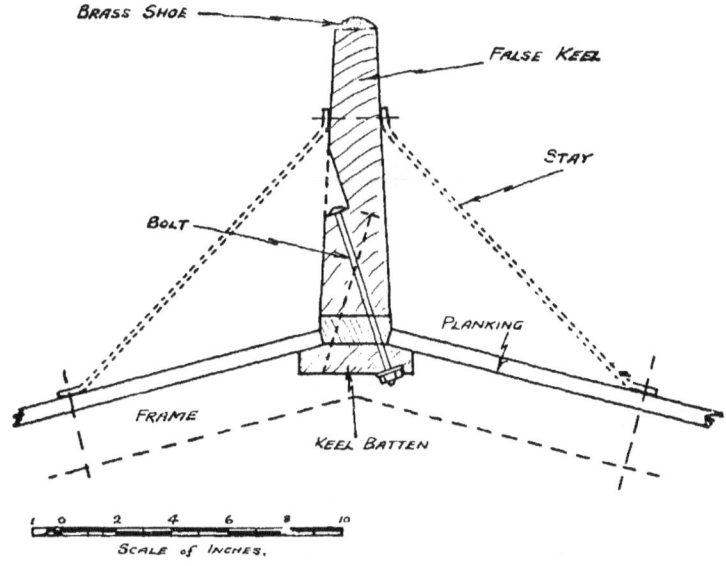

Fig. 10.—Section of false keel.

Where this fin is at its deepest, two side stays (shown in Fig. 10) may be fitted to give extra rigidity, but are not really necessary. The false keel can be dispensed with completely for still waters, but for sea or river use it

ERECTING AND PLANKING

steadies a hull of this kind considerably, and simplifies steering.

. . . .

The shaping and fitting of the chines varies but little from the keel procedure. One difference in lofting the sections occurs owing to the fact that the outer chines taper slightly towards the stem. The chine batten may also be made with a corresponding taper, and not only do such tapers ensure a neat finished appearance, but they also simplify bending. For our prototype, the chine batten is $2\frac{1}{2}$ in. wide by 1 in. thick, while the outer chine is $1\frac{3}{4}$ in. wide, tapering to $\frac{3}{4}$ in. from frame No. 2 to the stem, and 1 in. thick, tapering to $\frac{1}{2}$ in. at the stem. When lofting the chine sections, a separate tracing must be used at frame No. 1 and at the stem, with the reduced sizes of timber drawn on it. There is no harm in making the chines of more than one length of timber. Butt joints should be used, and in the chine batten the two parts may be well screwed to an oak block. The joint should be made on a fairly straight part of the chine, or distortion will invariably occur due to the increased stiffness at the joint. The outer chines are fastened at close intervals, so joints will have little effect upon the curve. The same rules apply to the keel members. Although the outer chine may be rounded-off completely after planking, the majority of this planing should be done on the bench, and great care must be taken to countersink the screws to a sufficient depth to allow for stopping the holes after the chine has been smoothed down. Many hard chine hulls, especially the larger ones, are made with the outer chine protruding beyond the planking, to form a ledge or rubbing strake. This is bad practice on a small boat, especially for sea or river use. Finish the chine as smoothly as possible in all cases.

PROPELLER SHAFT BORING

Chine fastenings are two 1¾-in. No. 10 brass screws through the batten into the frames, and one 1½-in. No. 10 brass screw every 6 in. along the outer chine. The stem fastening is quite simple and is sketched in Fig. 11.

The fitting of chines and keel will stiffen the hull considerably, and the temporary longitudinal ribbands may now be removed.

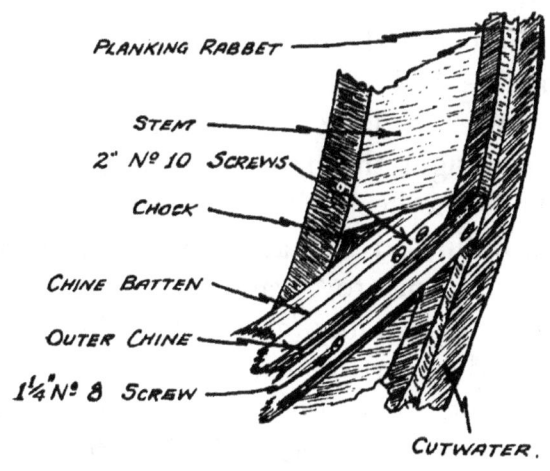

Fig. 11.—The stem fastening.

Propeller Shaft Boring

Although boring for the propeller shaft is often one of the final operations in building a hull, in our present case this process will be simplified considerably if carried out before the bottom is planked. The mechanical installation will be treated later, but for the present task it will suffice to say that the 1-in. diameter propeller shaft runs in a bearing and gland bolted inside to the shaft chock (shown in Fig. 12 (*a*)), and extends to the external bearing (at the propeller "A" bracket) without any protection, such as a stern-tube. Therefore, a plain hole is required, of 1⅛-in. diameter, bored obliquely through the keel and shaft chock. All angles and measurements may be taken

ERECTING AND PLANKING

accurately from the drawings, so that alignment of boring tools is fairly simple. There are two good methods of boring such a hole:

(1) By means of a boring bar.
(2) Using a specially lengthened bullnosed ship auger.

In either case the shaft chock should first be bolted into position.

The first method is the best and most accurate one, but it is only possible if a boring bar can be borrowed or made. An auger about $\frac{1}{4}$ in. larger in diameter than the boring bar is first run right through, as accurately as possible, starting outside the keel. The auger used should, preferably, be a bullnosed ship auger which, having no lead screw, has no tendency to " wander." The auger should be lengthened considerably (3 or 4 ft. added) by a blacksmith, and lined up in wooden chocks to ensure a fairly correct alignment. The boring bar, which consists of a length of about $\frac{5}{8}$-in. steel rod, threaded for half it's length with a cutter fixed through it at its centre, is then mounted through the auger hole. Its two bearings are a nut-plate, screwed to the inboard face of the shaft chock and a support bracket fixed above one of the frames and over the keel. If the primary auger hole did not prove accurate at its exit the nut-plate is set exactly correct, within the $\frac{1}{4}$-in. margin allowed, so that the final hole will be true. Should the auger hole be more than $\frac{1}{4}$ in. in error, the boring bar will have to be run through at a smaller cutter-setting first, to drift the hole over sufficiently to allow a central alignment of the nut-plate. A handle is attached to the external end of the boring bar, so that it feeds smoothly on turning. The arrangement is shown in operation in Fig. 12 (b).

The second method is certainly simpler to perform, but may not prove entirely successful. An auger of the

PROPELLER SHAFT BORING

full size, lengthened and lined up as previously described, is run right through. The auger is started by gouging a recess for it in the outer keel. If one does not care to risk this method and a boring bar cannot be obtained, a smaller auger, say ¾ in., could be used and the hole

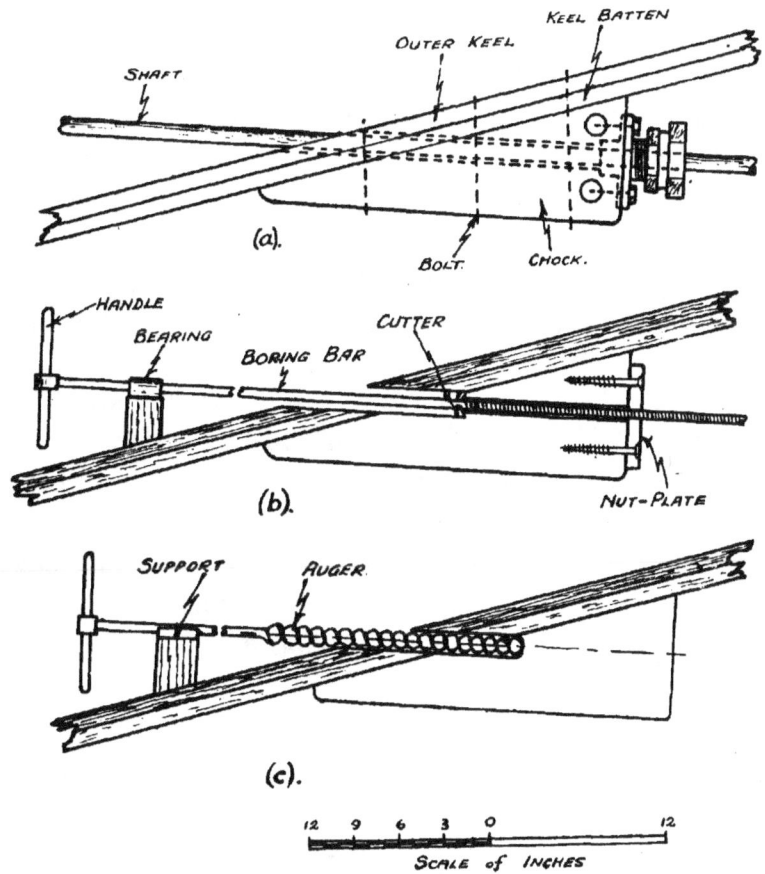

Fig. 12.—Boring for the propeller shaft.

enlarged by means of a red-hot bar; even professionals sometimes do this!

The inside of the boring should be given about ten coats of very thin white lead paint and the usual luting is essential under the shaft chock, where perfect fitting surfaces should be obtained. Galvanized carriage bolts

ERECTING AND PLANKING

are best for securing the chock, and oak is the best timber for this member. The fixing of the gland bearing depends on the particular fitting, but although coach-screws are generally recommended for these bearings, bolts are better, and they can be fitted as shown in Fig. 12 (*a*), with holes through the chock to take the nuts.

PLANKING

All builders agree that planking is the most difficult task for the amateur. However, if the correct method is adopted, it becomes rather more a tedious job than a difficult one. In a hard chine boat the bottom is planked completely before the topsides are started, and the garboard strakes (or planks adjacent to the keel) are fitted first.

There are four main systems used for planking hard chine hulls. All give a flush finish to the surface, so that except for the double-diagonal system, they are all carvel planking. Carvel planking may be defined as a skin composed of longitudinally laid boards with all edges butting to form a lamina of uniform thickness. Three different types of seam may be adopted:

(*a*) Caulked seams.
(*b*) Seam-batten caulked.
(*c*) Seam-batten uncaulked.

The sections in Fig. 13 show how each type of seam is formed, and Fig. 13 (*d*) represents the way in which double-diagonal planking is laid. More will be said of these various methods later, but for the present only caulked seam-batten planking will be treated. In this system a batten of pine, about $1\frac{1}{2}$ in. wide and $\frac{3}{8}$ in. thick, is placed behind every seam, each one running the length of the boat and notched into each frame to be flush with it. The plank edges are then fastened to the seam battens

PLANKING

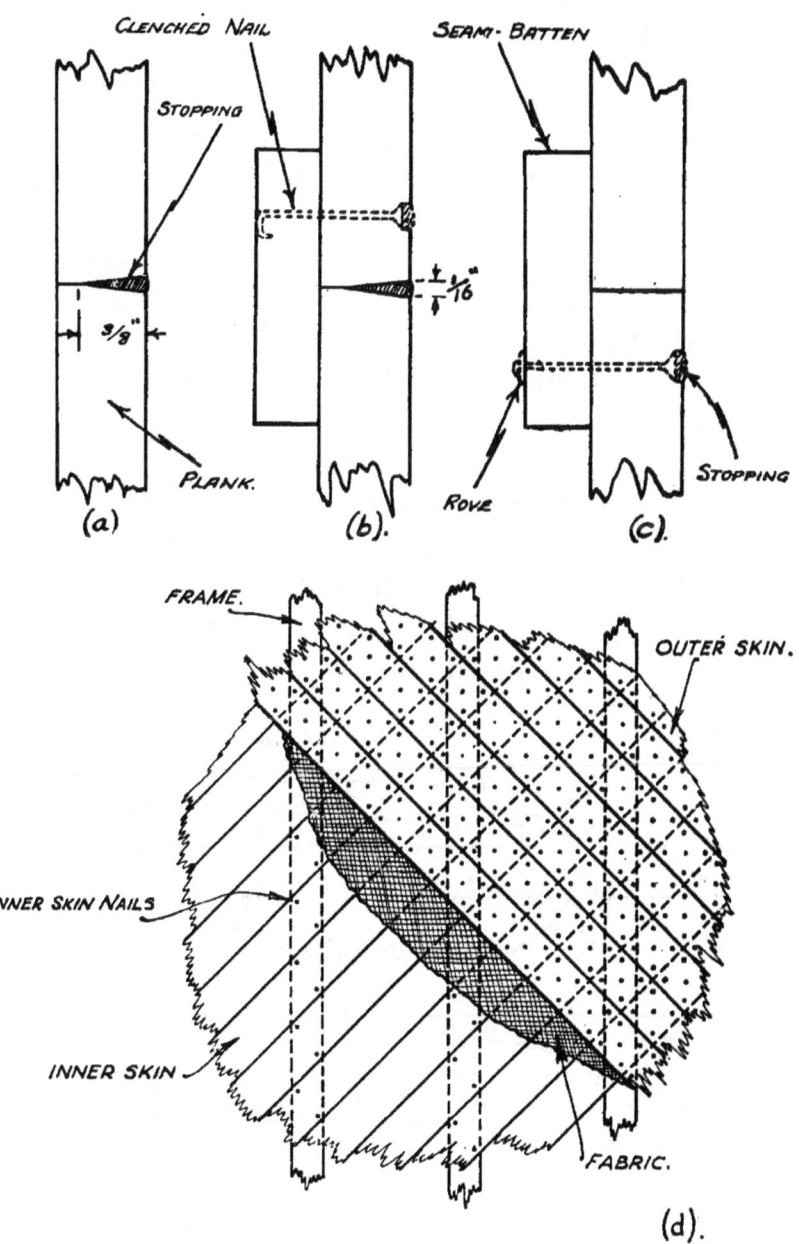

Fig. 13.—Carvel and double-diagonal planking.

ERECTING AND PLANKING

as well as to the frames, so that perfectly tight joints are obtained.

. . . .

To commence planking we must start with the two garboard strakes. Take a batten of springy wood, about 4 in. wide and ¼ in. thick, and lay it parallel to the keel, so that the edge which is towards the keel represents the position of that edge of the garboard strake which is farthest from the keel. The width of the timber to be used for planking must be known, so that this width can be set between the keel and the batten (usually called the spiling or offset batten), approximately amidships, or at such a position that distances to each side are less than the maximum plank width. The spiling batten can then be nailed or clamped in position, the end being cut roughly to the shape of the rabbet and tacked with wire nails into the rabbet as shown in Fig. 14 (*a*). The wire nails

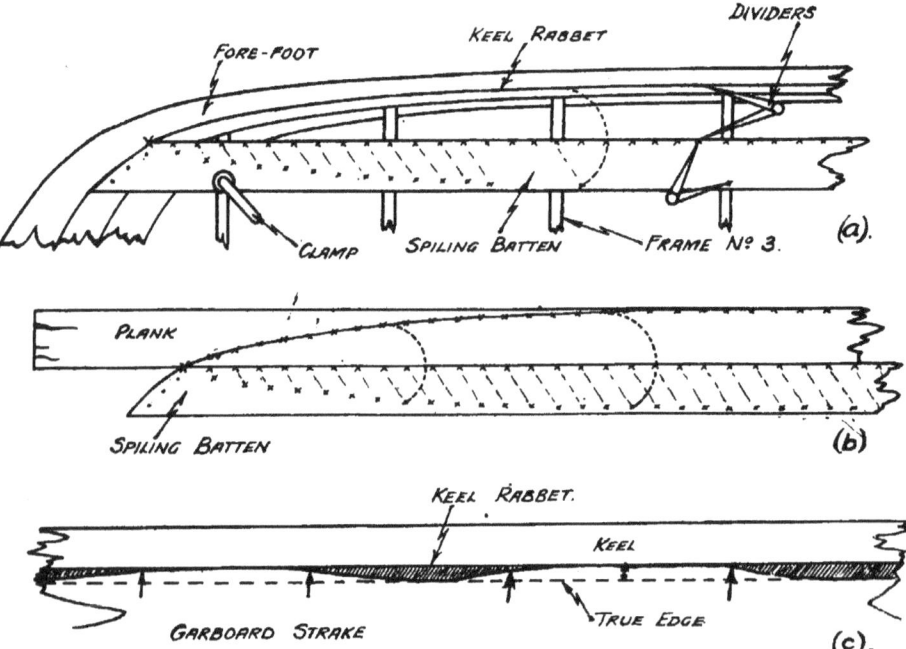

FIG. 14.—Method of marking garboard planks for cutting.

PLANKING

should not be driven right home, so that they can be withdrawn easily with pincers or a claw hammer. Divide the edge of the spiling batten into a number of equal spaces by making pencil marks about every 3 in. along it. Take a pair of dividers, and with one point on the first batten mark, open the legs until the other point rests in the keel rabbet. Swing this point around to somewhere close to the lower edge of the batten, and make a mark there. Join the final positions of the divider points by a pencil line, so that they do not get lost. Repeat this process at every 3-in. point along the edge of the batten which will appear as in Fig. 14 (*a*). Then remove the spiling batten and lay its divided edge along one edge of the garboard plank. Again, with the dividers, transfer all the distances from the batten to their respective points on the garboard plank. Join all these points, and a curve is obtained which should on cutting and planing allow the plank to fit snugly into the keel rabbet; see Fig. 14 (*b*).

This curve may be cut quite well on a circular saw, by a series of slightly curved cuts, although a band-saw will produce a cleaner and more accurate cut. If neither of these machines is available, a hand compass-saw or frame-saw is equally successful, if slower; a smoothing-plane will finish it to the line. If a perfect fit is not obtained, the plank must be clamped in place to observe and mark the "high-spots." The simplest way to do this is by making a pencil mark at each end of each "ridge," and another mark in the centre, to show the approximate amount to be planed off. Fig. 14 (*c*) shows an instance of this marking. Finally, the caulking bevel must be put on both edges of the plank. The caulking opening should be $\frac{1}{16}$ in. wide and $\frac{3}{8}$ in. deep, so that $\frac{1}{32}$ in. must be taken off each plank to a depth of $\frac{3}{8}$ in. A pencil mark across each frame, where the outer edge of the garboard strake comes, will mark the centre of the

ERECTING AND PLANKING

first seam batten. With a sample piece of the batten, the outlines of all the notches can be marked on the sides of the frames, and the ends cut with a tenon-saw. The chisel cuts should be taken from either end of the notch (up and down the run of the frame), using the end of the notch as a guide to prevent the chisel from cutting too deeply, and finally with a wide firmer chisel across the grain.

All planks must be laid in pairs, and matters are much simplified if the other garboard strake can be marked directly from the finished one. However, the keel rabbets may not be the same on both sides, so reverse the plank and see how it fits along the opposite rabbet. If the fit is a bad one, the spiling batten must be used again. In our particular example, and using $\frac{1}{2}$-in. pine planking, 4 in. wide, the garboard strake should not require steaming to enable its for'ard end to be strained into the forefoot rabbet. Steaming will be needed on later planks, however, so details of the required apparatus will be given shortly.

The seam battens are fastened to each frame with one 1-in. No. 8 brass screw. Liquid marine glue (such as Jeffery's "Hydro"), is spread thickly everywhere the planks seat, thus ensuring leak-proof planking, even without the use of caulking. The planks are fastened with three $1\frac{1}{4}$-in. No. 10 brass screws at each end and at intervals of 5 in. along the keel rabbet and the chine. The frame fastenings may be three fairly stout galvanized steel boat nails at each frame, or two nails and one screw. (Copper boat nails do not hold well when driven "dead" unless they are twisted.) Each plank edge is fastened to the seam batten with three $1\frac{1}{4}$-in. copper boat nails between each frame. These nails may be clenched or roved on the inside, the former being the neater and easier method. The nail is bored for, using a hand-drill

Plate 5.—The bottom planked and sheer battens fixed.

Plate 6.—The seam battens are notched into each frame.

ERECTING AND PLANKING

with a morse drill of suitable size, and the head is countersunk about $\frac{1}{16}$ in. Then the nail is driven until the head is almost flush—the point should then protrude about $\frac{3}{8}$ in. The tip must now be bent with pliers and the whole protruding part of the nail hammered over to re-enter the batten at a slight angle to the grain. The nail is then driven to completion with a nail-punch, hammer and "hold-on," so that the resultant fastening is a very strong one. Using roves, the nail is driven similarly until flush, then the rove is hammered on to it (concave side down) with a hollow punch (rove punch). The point of the nail is then clipped off, with end-cutters or side-cutters, leaving $\frac{1}{16}$ in. beyond the rove, which is riveted by means of the hold-on, nail punch and a light hammer (ball- or chisel-paned). These two fastenings are shown in Fig. 13. Further details of clenching and roving are given on page 134 and in Fig. 45.

Having laid the garboard strakes, the widest place between this plank and the chine must be found, and the edges of all the remaining bottom planks marked along the frame nearest to this place. Plank widths should not exceed 5 in., and are better at 4 in. Only good, clean timber, completely free from dead knots must be used. The drawings in Fig. 15 show the way a log is cut for slash-sawn (tangentially sawn) planks and two methods of cutting rift-sawn planks. Rift-sawn planks are the strongest for planking, decking or flooring and are least liable to warp or split. Slash-sawn planks bend more easily as the grain is more or less vertical (parallel to the surface), and may be used for the planks with the greatest curvature. These points should have been specified when ordering, but there is bound to be some variation, so choose the most suitable planks according to position and amount of curvature, using the best for the strakes below the waterline.

PLANKING

The spiling batten must then be clamped to the next mark on the widest frame, and the process of taking offsets repeated so that the next plank may be fitted without strain. It would be possible to clamp a plank into position without first shaping its edge, especially if it had first been steamed, but this practice should never be followed, as it imparts edge-set to the plank, which may split or open its seams even a long time after building.

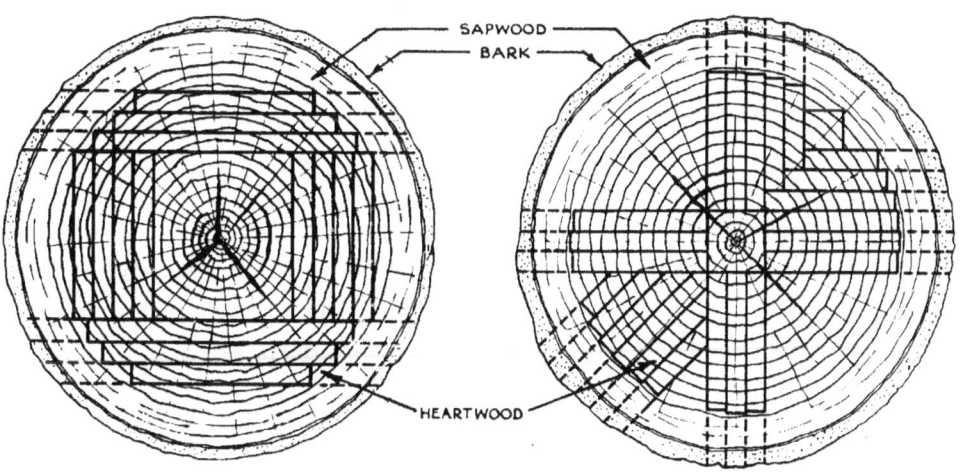

Fig. 15.—Methods of cutting planks.

A small amount of "edge-spring" is permissible in planking, but be careful not to overdo it.

The next two or three planks to the garboards will probably require steaming, as they twist through almost a right-angle towards the stem. The chine shutters, or planks next to the chines, will have both their ends terminating on the chine, but this is quite in order as long as the ends do not taper too finely, making secure fastening difficult. In this case the chine planks will

ERECTING AND PLANKING

have to be made "stealers," which are fitted into the adjacent plank in the manner shown in Fig. 16. Two lots of offsets must be made for the chine shutters, one row to get the curve of the adjacent plank, and one for

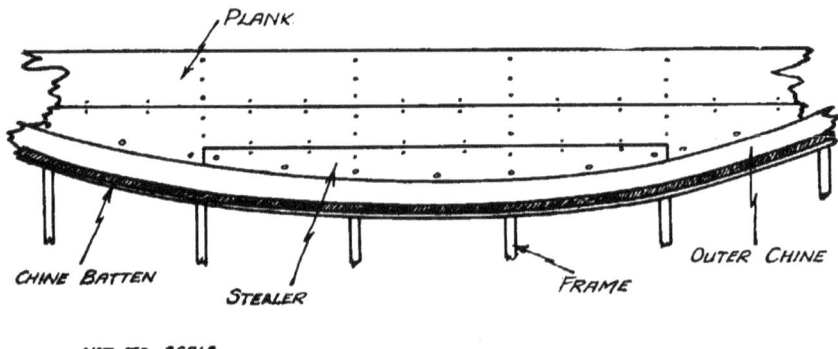

Fig. 16.—Planking at the chines.

the chine curve, making this plank a rather troublesome one to fit. Planks need not be in single lengths, but if butts are used, they must be well separated in adjacent planks, and they must be well screwed to an oak block on the inside.

CHAPTER IV

MISCELLANEOUS JOBS AND TOOLS

BESIDES normal carpenter's tools, a few extra implements are required in the boat-builder's kit, and although a list will be given later, unusual tools will be mentioned in the paragraphs concerning the tasks for which they are needed. The following notes should cover all special items required for planking.

The prospective builder naturally wonders "How many clamps shall I need?" A minimum number is difficult to state, because various-sized clamps are necessary; it is no good getting all big ones, and thinking that they will suffice for all positions simply because they close right up. Therefore, we must first invest in two small clamps, with a 3-in. opening and a " D " pattern frame. Then, three 6-in. " G " pattern and one 8-in. " G " clamp will be necessary.

These are the very minimum needed for the boat we are dealing with; if more can be obtained so much the better, and one or two adjustable clamps opening to about 2 ft. are almost essential. However, such implements are expensive, and unless they can be borrowed, the ingenious amateur will prefer to make a few extra clamps in one of the following ways.

Both bought and improvised clamps are pictured in Fig. 17. A simple wooden clamp, as shown in Fig. 17 (d) may be made very easily. The jaws are two pieces of 2-in. by $1\frac{1}{2}$-in. oak, about 18 in. long, with a slot morticed in the centre of each to take a section of 1-in. by $\frac{3}{16}$-in. mild steel bar, having several $\frac{1}{4}$-in. holes along it, and a steel pin above and below the jaws. Suitable wedges applied to the backs of the jaws enable any piece of work

MISCELLANEOUS JOBS AND TOOLS

within range of the holes in the steel bar to be clamped up tight.

Numerous variations of this style of clamp can be conceived, the one in Fig. 17 (*e*) providing a somewhat quicker device to set-up. The jaws are the same, but flat wood packing is used in the place of wedges and

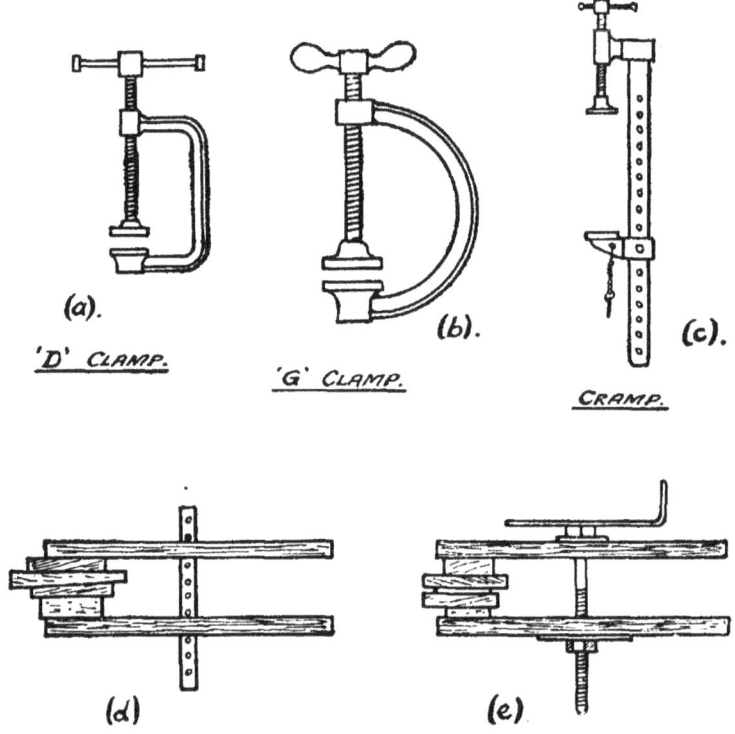

Fig. 17.—Clamps and cramps.

tension is applied by means of a bolt in the centre. A ½-in. bolt with hexagonal head and nut would do, and operation is simplified if the nut is welded to a piece of steel plate and screwed to the jaw, while the head seats on to a washer on the other jaw and has a handle of ⅜-in. steel rod welded to it, as shown in the drawing. The bolt hole in the upper jaw should be elongated to

MISCELLANEOUS JOBS AND TOOLS

allow the jaws to tilt from the parallel if required. The same applies to the mortices in Fig. 17 (d).

Long cramps are very handy for reaching across to the edges of planks and thus closing the seams completely before fastening. However, long cramps (as in Fig. 17 (c)) tend to get in the way while driving fastenings, so a better plan is to clamp or nail a block of wood to the frame concerned, and drive a wedge between this and the plank edge until the seam on the opposite side of the plank is closed, as shown in Fig. 18.

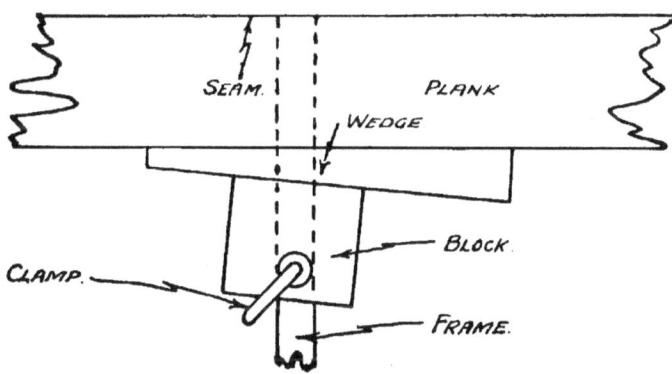

FIG. 18.—How to tighten a plank seam.

Steaming is necessary on certain planks in almost any boat. This is not to be feared, however, and having fixed up a steaming-plant, work will be simplified by using it on many of the planks which could be strained around dry. Normally, the steaming-box for our small chine hull need be only about 8 ft. long. It consists merely of two planks, spaced about 2 in. apart by side battens, to which the planks are nailed, and the width should be about 2 in. wider than the widest plank to be steamed. A section of the box is shown in Fig. 19 (a). One end is sealed, with the steam-pipe coming through it, and the whole box should be mounted on trestles to about waist-height.

MISCELLANEOUS JOBS AND TOOLS

The boiler should not be smaller than a 2-gallon oil drum, so that it may be left alone for a long period without attention, and the fire can be a gas ring-burner if indoors or a wood fire if outdoors. The steam-pipe should lead direct to the steaming-box from the top of the boiler, and can be brazed or welded there, the fixing depending on the type of can used.

When steam is flowing freely, the plank is inserted so that it almost reaches the end of the box, and the open

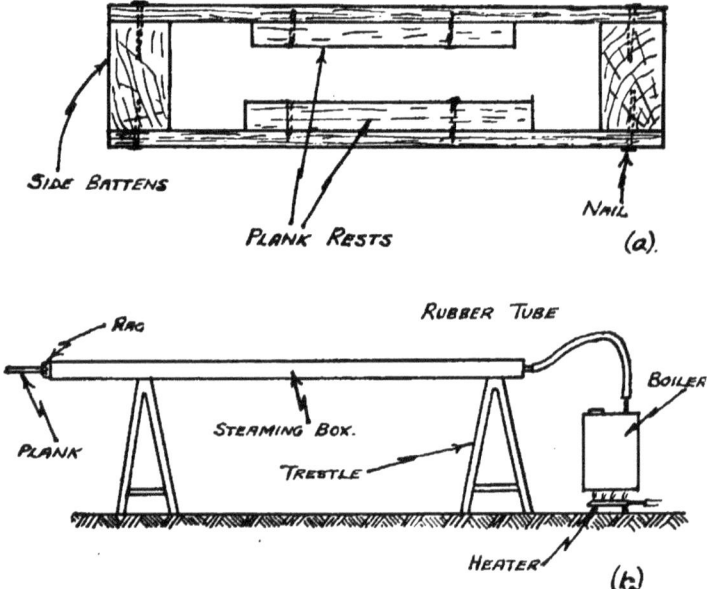

Fig. 19.—The steaming-box.

end is stuffed with cotton waste or rags. Interior support is aided if a number of wooden strips are nailed to the floor and roof of the box, in which case the plank should be moved frequently during steaming to allow the whole surface to be reached.

A typical arrangement is shown in Fig. 19 (b). It would be possible to set the steaming-box on edge, but then the intensity of steaming might not be even over the surfaces of the plank.

OTHER TYPES OF PLANKING

A $\frac{1}{2}$-in pine plank, about 5 in. wide, needs some two hours' steaming to become fully saturated, but where the bend is slight, considerably less steaming would suffice. Wherever possible a plank should be steamed only once, that is, after shaping its edge, it should be steamed and placed in position to mark any adjustments to the seam, but when replacing it for fastening it should have set to shape and should not require further steaming. Therefore, the first steaming must be a thorough one, and the clamps must not be removed for about six hours, preferably longer, depending on the weather and temperature. If fastenings are driven while the plank is damp, they will not bond well, and the heads will sink too deeply.

As an alternative to steaming, in cases where the design of a hull does not warrant the building of a steaming-plant, or where the preparation of getting up steam is too troublesome for one small job, an almost similar effect may be obtained by keeping the plank surrounded by rags well drenched with boiling water, for about half an hour. Constant attention is necessary, because the rags have to be repeatedly soaked in the hot water to keep the plank as hot as possible, until the timber becomes sufficiently supple. Heat is the important factor, so that whatever process is used, the plank must be positioned and clamped-up with the least possible delay.

OTHER TYPES OF PLANKING

Seam-batten planking has now been treated in detail, so before leaving the subject, a few points about the other systems are called for.

In the last chapter three other methods were mentioned, and drawings of them were given. The only difference between the method so far treated and the uncaulked seam-batten construction is that the former is provided with V-shaped seams to take stopping, and thus ensure a smooth finish to the hull, while the latter is only used for very high-class boats with varnished topsides, having accurately made seams that are laid tight all along, and have no space for stopping. Theoretically, a seam-batten boat should not leak even if her seams do open, but by caulking or close-fitting the seams, we can make doubly sure and produce a smooth surface.

Normal carvel planking (without seam battens) should be slightly thicker than its equivalent using battens, and with a caulked seam the seams just close tight on the inside and open in a " V " from a depth of about $\frac{3}{8}$ in. to the surface, with an opening of $\frac{1}{16}$ in. to the outside. Obviously, then, this type of planking must be far more accurately laid than any seam-batten system.

Certain large carvel-built yachts with varnished mahogany or teak topsides have close-fitting uncaulked seams without seam battens, but for an amateur's first attempt he is advised to build a hard chine hull, so that he can adopt seam-batten planking. In all these methods the actual planks are fitted just as already described; the only difference is in the seams.

Double-diagonal planking is entirely different, and some readers may prefer it to any of the other methods. It consists of two separate skins of planking with a layer of waterproofed fabric between them, the inner layer being formed of short lengths of narrow planking nailed to the frames at an angle of 45 deg. to them and the outer skin of similar strips, nailed to the frames and clench-nailed to the inner skin, sloping the opposite way at 45 deg., so that the two skins meet at 90 deg.

This system is especially suited to hard chine boats, although it has been used with great success on round bilge hulls. It is naturally very economical of timber, as no fine long lengths have to be picked, and the inner

OTHER TYPES OF PLANKING

skin need not be first-class timber. There is little shaping to be done to the planks, so this method is practically a carpenter's job. For these reasons most of the small boats built for the Admiralty during the war were double-diagonal planked.

To the experienced boat builder, planking takes longer to lay this way, and it is certainly not so rigid as carvel. The amateur will find it somewhat quicker than carvel, but on some of the curves near the bow he will find it rather tricky, and his inner skin when completed will probably appear a very bad shape. When the outer skin is laid and fastened to the inner skin, the shape is corrected and the result will be pleasing.

A disadvantage is that this method uses more fastenings than any other, and as the resulting thickness of the whole is also somewhat greater than carvel, the finished weight is greater. Another drawback is in the matter of repair. It means a tremendous amount of stripping to patch up a hole or in the event of rot along the keel, and the result can never be as strong as the original job.

The inner skin can be as thin as $\frac{1}{4}$-in. pine, but for the hull we are treating $\frac{5}{16}$-in. would be better. The planks should be between 2 in. and 3 in. wide, nailed to the frames with $\frac{3}{4}$-in. copper boat nails, and their edges should all be planed to fit to their adjacent planks. The inner skin nails should be driven at a slight angle, to tend to draw the joint together, and into the *edge* of the frame (say an $\frac{1}{8}$ or $\frac{1}{4}$ in. in from the edge), so as to have plenty of timber as the nail drives on, and to be out of the way of the outer skin fastenings. Where the planks take a curve, it will be found that perfectly straight edges will not always meet up properly, and although this skin does not have to be water-tight, all seams must be close, and plank edges well painted with a thick mixture of white lead and linseed oil before nailing.

MISCELLANEOUS JOBS AND TOOLS

The fabric layer (usually unbleached calico) is best laid in strips, running in the same direction as the outer skin seams, with the overlaps always coming in the centre of a plank. Each strip of fabric is laid on a thick coating of whitelead paint or liquid marine glue, spread over the inner skin, and only one strip should be laid at a time, so that the outer skin can be laid immediately. More paint or glue is spread over the canvas before laying the outer skin planks. While the fabric is still soft with the paint or glue, the fastenings should go in as indicated in Fig. 13 (d). Each nail should be just a little more than the thickness of the planking (outer) from the edge of the plank.

The outer skin should be of about the same scantlings as the inner skin, and must be fitted even more precisely, so that no caulking is necessary, and that the hull shall appear perfectly smooth on finishing. The frame fastenings must be placed to avoid those already driven, as shown in the drawing in the last chapter. All these outside fastenings should be countersunk at least $\frac{3}{32}$ in. to provide room for stopping, though a stronger job is to leave them flush with the surface.

Finishing

Before caulking, stopping and painting or varnishing woodwork, the surface should be prepared by planing, scraping and glasspapering, or the finished job will not be satisfactory.

Planking or decking joints should first be " traversed " with a smoothing plane, i.e. planed across the grain of the wood, usually at an angle of about 45 deg. to the grain. This can only be done where the surface is flat or convex. The concave for'ard sections of the topsides are difficult to plane. A " circular " plane (Stanley No. 20), if obtainable, will traverse hollow places and a

CAULKING AND STOPPING

small block plane (Stanley No. 102) will plane in the direction of the seams if the blade is set (on the oilstone) with a slightly curved edge.

Planing is followed by scraping, either with a cabinet scraper or one of the handled variety, such as the " Scarsten." Most professionals prefer the cabinet scraper, which is a plain piece of steel, and is certainly the most effective if one has learnt the trick of sharpening it. Hold on edge or screw in a vice and stroke along the edge with a cutlery steel or the rounded side of a gouge. The steel should be applied with heavy pressure at a slight angle to the edge and produces a sharp burr. If unsuccessful, get a carpenter or shipwright to demonstrate.

Finally glasspaper across and then along the grain. A cork block is generally used to hold the glasspaper.

Caulking and Stopping

Before turning the boat over to plank the topsides, several other jobs must be done to the bottom, as well as to other regions, and caulking will be much more easily done whilst the bottom is uppermost. The seams must first be thoroughly cleared of shavings and sawdust.

Caulking is strictly defined as the laying of oakum or cotton into the seams, and paying means the application of pitch or putty to the oakum or cotton to keep the latter in place and to keep it as dry as possible. However, in a small boat, caulking usually means stopping the seams with a type of putty, and in a seam-batten boat there is no object in using oakum at all. Therefore, all we need is the proper putty stopping and a fairly springy-bladed glazier's putty knife.

The seams must first be very well painted with a thin white lead priming paint, so that the raw wood cannot absorb the oil from the putty. This paint may be applied with a flat seam-brush, a small sash-brush or an old tooth-brush. It must cover all places inside the seams.

The best possible putty for the seams of a painted hull is made by intimately mixing 1 part of white lead (ground in oil) with 2 parts of whiting and $2\frac{1}{2}$ parts of red lead (powder), by weight, and beating the mixture with a little raw linseed oil until a soft putty is obtained. There is an art in preparing this stuff, but it requires only one or two mixes to learn it. The white lead should be broken into many small lumps, and this and the powders thoroughly mixed with the hands or a putty knife. Very little linseed oil is required. As the oil is added, the whole mass is beaten continually with a mallet or old rolling-pin, covered with whiting powder to prevent the putty sticking to it. When nearing the point of a homogeneous mix, the putty should be beaten very hard, whence it will become extremely soft, of a creamy texture, meaning that too much oil has been used. The only way to correct this tendency is to leave a little of the powder mix, which can now be added to the remainder until a sufficiently stiff mix is obtained.

This putty never quite hardens up, so that small movements of the seams will not crack it or spew it out. The knack of working the putty right into the groove of the seam will soon be acquired, and the surface should hump slightly, above the level of the planking, because on drying this stopping shrinks somewhat and will leave a depression if the surface is made flat during paying.

No stopping should be left on the planks at each side of the seam, and if some is left there it will have to be scraped off, thus also scraping the surface of the seam stopping and rendering it less secure on drying. The slight smoothing down which will be needed should be done with very fine glasspaper, preferably used with water (using waterproof glasspaper) and then only

PAINTING

after the stopping has had about a week in which to harden.

Some boat-building yards use "caulking guns" to apply stopping to the seams. These work on the same principle as a car grease-gun, and are fitted with an elongated nozzle and adjustable support which rests on the planking, so that the stopping may be forced right into the seam as the gun is moved along it.

Where plain seams, without seam battens, are used, it is usual to lay a few strands of caulking cotton in the seams before stopping. This caulking does not serve the same purpose as the oakum in the seams of a bigger craft. It merely takes up any openings in the inside of the seam (which should be a tight butt) and serves as a key to hold the stopping in place and prevent it from being forced right through the seam.

Caulking cotton is like the yarns of a cotton rope, and about two strands of it should be twisted well between the fingers, fed into the seam in a series of fore-and-aft loops, and forced in with a very narrow caulking iron or the rounded edge of the blade of a putty knife, or, better still, by means of a tool known as a caulking wheel; a thin metal disc mounted on a handle. The cotton should be put in after the seam has been painted, but a brushing of raw linseed oil should be applied a day before stopping. Great care must be taken to see that the seams are entirely free from sawdust or shavings before caulking commences.

Painting

Before painting the bottom, the false keel should be bolted permanently in place, on to a luting of white lead, and the rubbing band should be screwed to the stem and keel. In our present case this band is of $\frac{5}{8}$-in. half-round brass, and it should be bent to shape round the stem before the screw holes are drilled in it. These holes

MISCELLANEOUS JOBS AND TOOLS

need to be at about 4-in. intervals, and bored to take the 1-in. No. 8 brass screws, countersunk flush with the band, the screws being filed off smooth after driving.

Paint the wood first and let it dry before affixing the band. The band should terminate at the after end of the false keel, but an inch of spare should be left above the stemhead, to be trimmed off when the deck level is determined. There is no reason why a galvanized iron rubbing band should not be used, but this section stuff is not readily obtainable galvanized.

The bottom painting is very important, and unless properly done, it will mean a complete scraping within a year. The priming coats are the most important. For best results only a really reliable make of paint, or one mixed by the user, should be employed. Details of the recipes and methods of mixing such a paint will be given later on; it remains only to say that two priming coats of very thin paint (with as much turpentine as linseed oil) are needed, followed by about five coats of a less volatile paint (with very little turpentine but a small amount of driers) containing the same weight of red lead as of white lead. Each of these five coats should be rubbed down with a medium waterproof glasspaper and water, or a pumice-composition block and water. Pumice powder on a pad is too fine for this purpose. Each coat must be fully hardened before rubbing down.

Lastly, the final coats may be applied. Enamel is best for the finishing coat, but a hard gloss paint will do; in either case the penultimate coat should be the undercoating recommended by the makers. The best colours to use are red, green, brown or black. Before painting or varnishing dinghies, a good idea is to put a couple of gallons of linseed oil into the bilges, and keep this moving about inside for several days until the planking becomes saturated with it. This waterproofing and rot-preventing

TOPSIDES PLANKING

measure may be applied to bigger boats by giving the planking several brushings of raw linseed oil before painting.

If the boat is to be kept long in salt water the process described above should not be followed for the part which will be below the waterline, which should instead receive two coats of a copper or antifouling paint. In this case do not use any other paint but apply the special paint to the bare wood.

TOPSIDES PLANKING

Before turning the boat over, we must fit the sheer battens, the sheer planks, and preferably the next planks to these as well, so that the hull is sufficiently rigid to withstand the strains of lifting. The sheer batten is a

Plate 7.—The sheer batten in position.

MISCELLANEOUS JOBS AND TOOLS

single length of 2-in. by 1-in. pine, notched into the frames so that it comes exactly in line with the sheer marks, as shown in Fig. 20. One of these battens is fixed to each side of the hull, and is screwed to the stem and transom frame and to each frame with two 2-in. No. 10 screws.

Where the frames have flare (curve outwards towards the sheer) for'ard of amidships, the outboard corner of the batten must be made in line with the sheer mark; where the frames have tumblehome (curve inwards) abaft amidships, the inboard corner must align, as shown

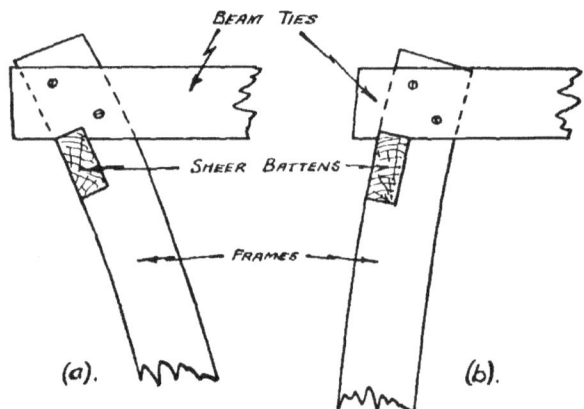

FIG. 20.—Positioning the beam ties.

in Figs. 20 (a) and (b) respectively. The sheer batten takes the place of the shelf and clamp found in a bigger boat, and provides a firm fixing for the top of the sheer plank. The sheer battens will undoubtedly require steaming.

The two sheer strakes are laid next. As our boat has a straight sheer, the sheer planks would have to be of an unreasonable width amidships to terminate at both the transom and the stem, so they are generally laid with a couple of inches width at the transom, the maximum width (about 5 in.) amidships, and then tapering rapidly

TOPSIDES PLANKING

to end on the sheer battens, a few feet from the stem. The foremost few feet will need steaming.

Offsets may either be obtained with a spiling batten and dividers, or by steaming the plank and clamping it in position in its full width, whence the sheer line can be marked direct to the plank by running a pencil along the sheer batten.

The only disadvantage of this last method is that the plank will be set to a curve on removal, making it awkward to cut and plane to shape. Obviously the plank will have to be shaped roughly along its sheer edge before steaming, or it will foul the beam ties on the frames, where they protrude to rest on the erection stringers. Alternatively, the beam ties can be notched to take the extra width of the plank, if they are deep enough to allow for this. The seam batten must terminate on the sheer batten, so to avoid a very long flimsy taper the seam batten should be notched into the sheer batten in the manner shown in Fig. 21. This drawing also shows the fastenings for the

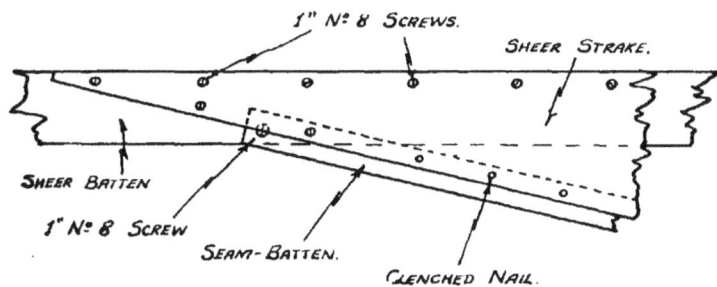

FIG. 21.—The seam batten is notched into the sheer batten.

for'ard end of the sheer strake. Its remaining fastenings are exactly as for the bottom planks, but 1-in. No. 8 brass screws are driven at intervals of about 5 in. into the sheer batten.

It may be possible to mark both sheer strakes from the one set of offsets on one spiling batten, but if the direct

MISCELLANEOUS JOBS AND TOOLS

marking method is used, it will have to be done separately for each plank. With both sheer strakes fitted the boat can be turned over, but a wiser plan is to get the next pair of planks laid first.

These strakes are adjacent to the sheer planks, and they should run the length of the hull this time to end on the stem. The spiling batten will have to be used here, and steaming may be needed. In order to get these two strakes to terminate on the stem, a maximum width of 6 in. is allowable for these strakes and the sheer planks. The remaining topsides planks should not be over 5 in. in width.

Turning Over

A cradle must be erected for the boat to rest on after she is inverted. This is best made in the form of two V-shaped bearers to fit the shape of the bottom, either to be cleated to the floor, or fixed to two longitudinal bearers resting on the floor. The former is easier for a timber floor, the latter for a concrete or earth floor.

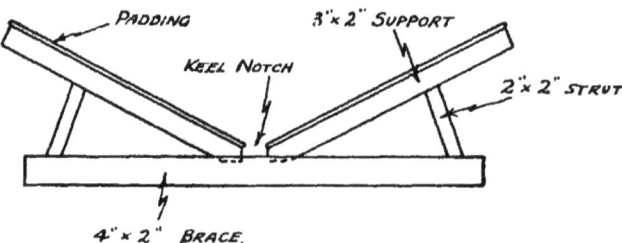

FIG. 22.—Make the cradles before turning over.

Make the hull supports as shown in Fig. 22 by fitting them to certain positions on the bottom (whilst it is uppermost), the best positions being just for'ard of the propeller shaft boring (below frame No. 6) and directly below frame No. 3.

These supports must each come in line with a frame

TURNING OVER

and should be well padded to protect the bottom enamel. The after support may be raised higher than the for'ard one to bring the waterline about level and to assist later with the fitting of the rudder and propeller shaft.

If convenient, place the cradles beside the hull and lift or roll it over. The boat should only weigh about 2 hundredweights so far, so four men can handle it without having to rest it in any awkward position during turning.

Many small jobs should now be done before the remaining topsides planks are laid, because the openings in the sides between the frames are very useful for fixing certain internal members, such as the engine bearers and rudder mechanism. The deck beams also can be fitted immediately and all the beam ties removed.

CHAPTER V

COMPLETING THE HARD CHINE RUNABOUT

IN the present stage of construction our hard chine hull is resting on a cradle, with her bottom completely planked and two strakes of topside planking adjacent to the sheer fitted on each side. The deck beams must next be considered, and when these are placed all the beam ties can be taken from the tops of the frames, imparting a more shipshape appearance to the hull.

A typical arrangement for the beams in our prototype is shown in Fig. 23. There are the chief beams, the

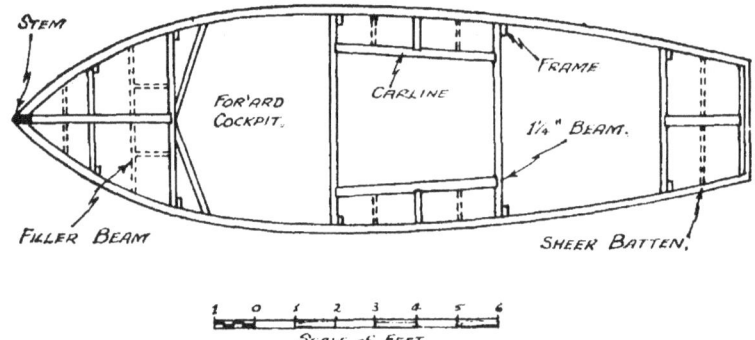

FIG. 23.—Deck beams and carlines.

minimum number possible, and they are all of oak, sided $1\frac{1}{4}$ in. and moulded $2\frac{1}{4}$ in. Any other good hardwood would do, and these members need not be natural crooks, with the grain following the curvature (or camber).

If crooks were used, less wastage would naturally result, but several beams can be cut from a single wide plank of straight oak and the cost would probably be less than for crooks. The use of a band-saw in cutting out also makes for less wastage of timber, as the cut can be kept accurately to the lines and two beams parted with a single cut.

Plate 8.—The hull nearly completed, with covering boards in place.

Plate 9.—A trial fitting of the engine hatch.

COMPLETING THE HARD CHINE RUNABOUT

A suitable camber for these beams is a central rise of 1 in. in 20 in. of span. Therefore, when the boat has a beam of 5 ft. her deck will arch to a height of 3 in. above the sheer lines. The first beam may be marked by trial and error, using a length of string as a radius arm, tied to a pencil at one end and pivoting on a nail or other anchorage at the opposite end. Having sawn and trued this beam with a spokeshave and plane, the remainder can be marked from it; the first beam to be sawn should, therefore, be the longest one.

The carlines of $1\frac{1}{2}$-in. by $1\frac{1}{2}$-in. oak are fitted at right angles to the beams, but except for those used to hinge the engine hatches, they serve no essential purpose. They do add to the general stiffness of the hull and are useful for the attachment of various interior fittings. They should always be notched into the frames and end-fastened with a single $2\frac{1}{2}$-in. No. 12 obliquely placed brass screw.

This deck framing, as pictured in Fig. 23, is satisfactory for a solid or double-skin varnished deck, but for the more popular light pine painted or canvased deck additional framing is required in the form of the extra " filler pieces " shown dotted in Fig. 23. This system will be discussed in more detail under a later paragraph concerning decks, because these extra pieces are often put in after the decks are laid.

Whenever possible the stout beams should be made to meet the sheer batten at the side of a frame so that they may be fastened to a frame as well as to the sheer batten. A typical section of the joint is shown in Fig. 24 (*a*). A shelf is not necessary using the scantlings which have been recommended so far, provided a wide covering board is fitted, but if the boat is to be built to a lighter specification a shelf-cum-clamp must be fitted, running the length of the hull, and about $1\frac{1}{2}$ in. square, as shown in Fig. 24 (*a*).

COMPLETING THE HARD CHINE RUNABOUT

This member will require steaming and should be a single length, but alternatively, the more conventional separate shelf and clamp can be used, as shown in Fig. 24 (*b*). In any case, do not dovetail or even notch the beams into the sheer batten. The boat is rigid enough without this and it would only weaken the sheer batten. Just use one $2\frac{1}{4}$-in. No. 12 brass screw through the frame and one $2\frac{1}{2}$-in. No. 12 screw through the sheer batten into the end grain of the beam.

Normally each beam tie can be completely removed before fitting a beam in its place, but a good precaution

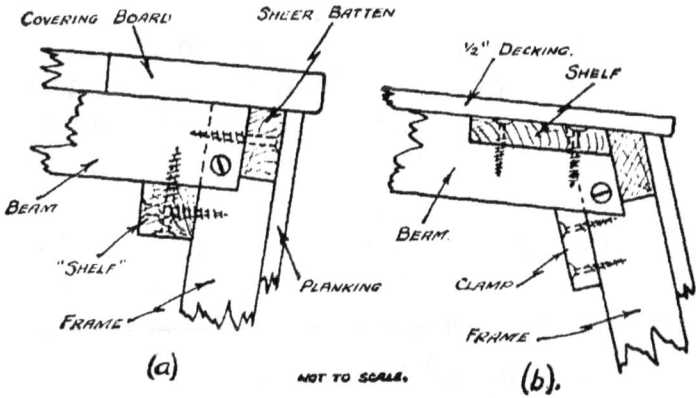

Fig. 24.—A shelf is not necessary for our prototype.

is to tack a temporary tie lower down the frame before removing the main one. Due to the changing angle of the sheer batten, the deck beams must be made to drop into position from above (viz. Fig. 24 (*a*)) where the frames have flare (for'ard of amidships) and to fit from below where the frames have tumblehome, aft, as in Fig. 24 (*b*).

The foremost one or two beams will have slight bevel, increasing to a maximum at their centres, and the central carline here will not be straight along its top, but will have a slight fore-and-aft curve, as well as an athwartships curve corresponding to the camber of the beams. The

COMPLETING THE HARD CHINE RUNABOUT

covering board will be mentioned fully in the section about decks, but if one is to be fitted it should be made and screwed down permanently now, although there is no serious objection to leaving it until just before decking.

Before leaving this subject the importance of using only well-seasoned and dry timber must be stressed. Should any expansion or contraction of the beams occur the deck-line of the boat may become warped, the frames will be strained and leaks started. The beams are so long amidships that this effect is magnified.

Engine Bearers

Having removed all the beam ties and replaced them by a fewer number of deck beams, we have, with the open sides and top, the maximum possible accessibility to the interior of the hull. The fitting of the engine bearers is therefore simpler now than later.

The motto with regard to these members is " as long and as stout as possible." By following this rule engine vibrations are damped and distributed over the maximum length of the ship, and the engine has little tendency to shift its position due to any straining or working of the hull. Such bearers also add considerably to the general strength of the hull.

It is customary and wisest to make the bearers of straight timber, so their length is governed by the fore-and-aft rise of the bottom frame futtocks and by the maximum permissible moulded depth of the bearers. For the hull under consideration two bearers are required at a spacing just greater than the maximum width of the engine, and each sided $2\frac{1}{2}$ in. and moulded 6 in. at their deepest points. The undersides of these members are joggled (notched) over each frame and fastened right through at each frame with a single $\frac{1}{4}$-in. galvanized bolt, fitted as Fig. 25 so that all the nuts are inside.

ENGINE BEARERS

Do not fit the bearers permanently yet. Mark out the shape of the underside of the bearers by placing them in correct alignment in the hull and, with a pair of dividers,

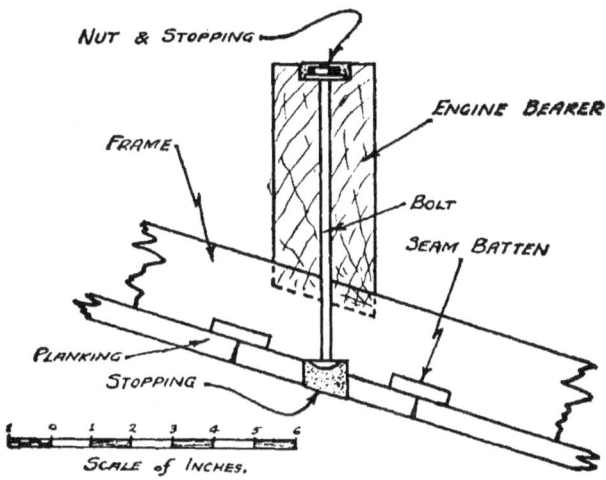

FIG. 25.—The engine bearers.

mark a point above each frame, on the bearer, so that in each case the distance from the point to the top of the frame is equal to the greatest distance that the underside of the bearer is away from any frame. The idea is

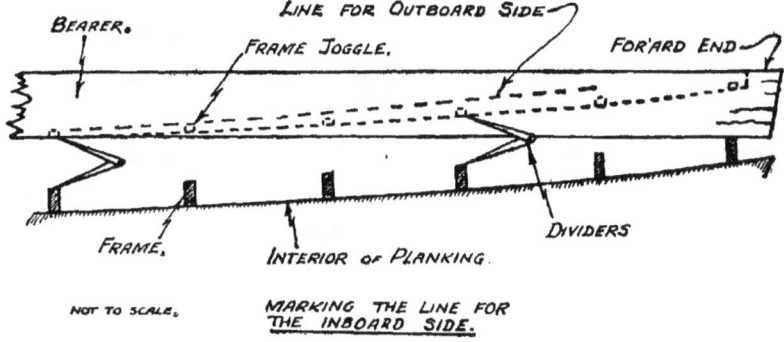

FIG. 26.—Shaping the underside of bearer.

sketched in Fig. 26. Do exactly the same to the other sides of the bearers, and join all these points to get the curve of the underside and the necessary bevel.

COMPLETING THE HARD CHINE RUNABOUT

On cutting along these lines the bearers should touch each frame when placed into the hull, but a few high or low spots do not matter, as the frame notches can be made accurately afterwards. Care must be taken that in marking the points with the dividers the opening of the dividers is such that after joggling the bearers $\frac{1}{2}$ in. at each frame there is a minimum depth of about 1 in. left at the extreme ends of the bearers. Using bearers of 6 in. moulded depth this means that the dividers must not be opened more than $4\frac{1}{2}$ in.

Even a canting table band-saw is not ideal for cutting out the bearers, because the bevel changes continually from a maximum at the for'ard ends to nil at the transom if they reach as far as this. The quickest method with the circular saw, will be a series of straight cuts to finish to the inboard lines on each bearer and then a further series of cuts, guiding by hand, to apply the bevel, and thus finish to the outboard lines as well. The surface can then be planed.

Position the bearers and mark each notch for the frames. Cut these notches, each $\frac{1}{2}$ in. deep, making any allowance for gaps between the bearers and frames due to bad fitting. Again position the bearers and drill all the bolt holes right through to the outside of the planking.

Before fastening the bearers permanently the whole interior of the hull should be given two coats of thin red lead paint, the first coat being the same mix as for pink priming; the second as for the final coats prescribed for the bottom planking. Work both coats very well into every crevice, not forgetting the limber holes and where the planking fits into the forefoot rabbet, but do not allow the paint to collect in pools or to leave air spaces in cracks.

Paint the bearers separately and then bolt them in. Lute the bolts well with thick red oxide paint before driving them in from below. Provide a ¼-in. countersink for the heads, so that they may be stopped over afterwards to prevent leaks. Lute well under the washers and nuts (which should also be countersunk for neatness, if only where they occur in the cockpits) and rivet up after tightening finally.

RUDDER MECHANISM

The mechanical parts of this boat will not be treated in detail, because they are a separate section from pure boat building, and very few readers will wish to build a boat similar to our prototype. However, the principles so far covered apply to all chine-built boats, the only item peculiar to power-driven craft being the engine bearers. Even this section is not useless to the sailing-boat builder, because it entails the principle of fitting a straight piece of timber into a curve, a task which occurs frequently in boat building.

The neatest rudder fitting for this runabout is the concealed gear shown in Fig. 27 (*a*). This has one main disadvantage. As the propeller requires to be as far aft as possible, to enable the angle of the shaft to be nearly horizontal, a rudder fixed inside the transom naturally limits this ideal propeller position. Furthermore, a balanced rudder, with its spindle fixed through its centre of gravity, is best for a runabout, but this would move the propeller even farther for'ard. Therefore, the alternative arrangement shown in Fig. 27 (*b*) may be adopted. It is simple to install, but may not be liked for its appearance.

Connexion to the steering wheel is commonly made by means of two stranded steel wire cables, with several turns around a drum on the steering wheel axle and actuating an arm attached to the head of the rudder stock. Wire is permissible for this small boat, but for a bigger vessel either rods or chain must be used. Carry the wires together around one side of the hull, passing through copper sleeves in the frames. Put rigging screws in the wires so that they may be tensioned and adjusted.

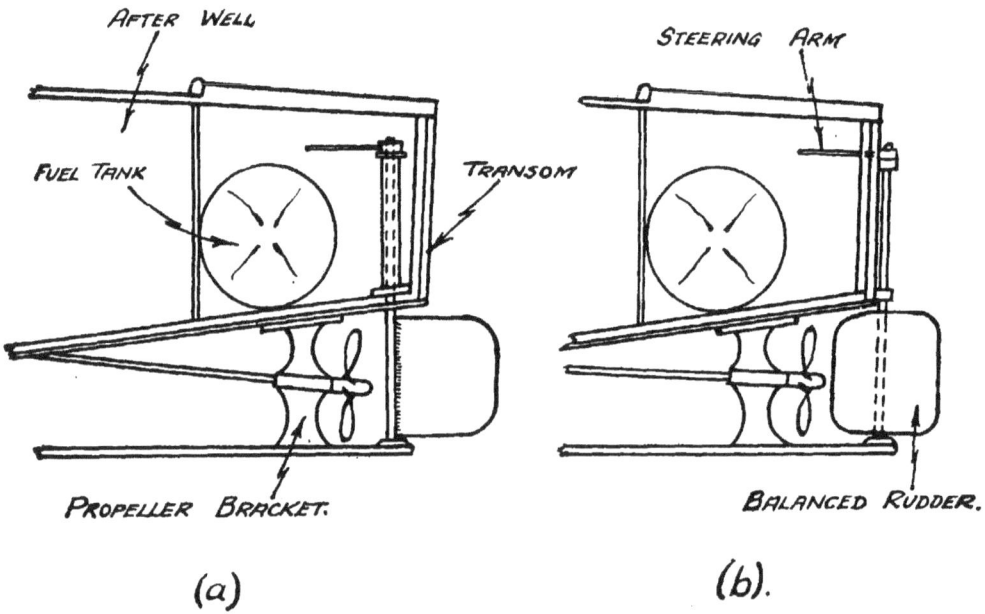

FIG. 27.—Alternative types of rudder.

A method of connecting the steering cable to the rudder by means of a link to the tiller arm is shown diagrammatically in Fig. 28. This enables the tension of the cable to be maintained in any position of the rudder and obviates back-lash. In this case a single cable is used, both ends being made off on the turnbuckle. A stout set-screw, say a $\frac{3}{8}$-in. B.S.F. bolt, should be fitted in the boss of the rudder arm so that the position of the blade may be adjusted to the movement of the arm. The stop-block is to prevent a lock-up, which would occur if the link and rudder arm were allowed to come into alignment.

The strands of steel cable should be soldered into a

MOTOR INSTALLATION

solid mass before cutting with a fine half-round file, to prevent the strands from unravelling. The cable, if not too large a diameter, may be made off with bicycle brake eye-clamps, cut off as shown in the sketch, the holes being enlarged to an oval shape, if necessary, with a small rat-tail file. For larger cables U-bolts or special clamps are used.

Motor Installation

The propeller shaft and its bearings are treated very fully in various books (such as *The Motor Boat Manual* and *The Marine Manual*) so details will not be given.

A reversing and feathering propeller is ideal for a runabout, either the Duerr or Wortham-Blake types proving suitable. These types, which incorporate a thrust bearing, are simple to install. The elimination of a reverse box lessens the weight of the engine as well as the cost, although a reduction gear may be necessary to maintain efficiency and keep the propeller speed below 1,000 r.p.m. A 14-in. reversing propeller and transmission costs about £10 to-day. Its only disadvantage is the comparatively high cost of repair should a blade or the shaft get damaged.

No matter what care is taken in mounting the engine, there is always the possibility that, upon trial, it will produce unpleasant vibrations, and, to obviate this and eliminate any danger due to the straining of the hull, rubber mountings and flexible couplings are strongly recommended.

Flexible rubber bush couplings for marine use are manufactured, in many sizes, by Silentbloc, Ltd., who also specialize in rubber mountings for engines. It must not be overlooked that short lengths of flexible tubing, such as are used for cars, should be inserted in the exhaust and petrol pipes; in fact, *all* connexions to the engine must be flexible.

COMPLETING THE HARD CHINE RUNABOUT

Install the engine as low as possible, and, even though a flexible coupling is used, it is still of great importance that lining up with the propeller shaft should be carried out with accuracy. Normally, the engine is coupled by flanges to the propeller shaft, and alignment is ensured by checking the coupling with feeler gauges and for the corresponding of the bolt holes in all positions.

Place the water intake skin connexion as close to the keel as convenience allows, and provide it with a good filter and inspection chamber. If possible, the spent circulation water should enter the exhaust pipe close to the engine so that the cooled exhaust pipe can be fitted close to the frames in its stretch from the engine to the transom. A side exhaust is unsatisfactory in an open boat. The petrol tank may be concealed under the after deck, with its filler cap on the king plank.

A good table concerning the sizes of motors and propellers for many types of hulls is given in *The Motor Boat Manual*.

Final Planking

If the bulkheads are to be of waterproof plywood or other sheeting they should be fitted temporarily now. Permanent bulkheads would obstruct the use of planking clamps at this stage, and tongued and grooved partitions are easily fitted after planking.

Mark all the remaining plank seams on to the frames so that the last pair of planks shall not be peculiar in shape. To keep to these marks both edges of the planks may have to be shaped, but this is no difficulty and well worth while for the sake of appearance. Work downwards in pairs so that the shutter planks come adjacent to the chines. Fastenings are exactly the same as for the bottom planking, and no steaming should prove necessary.

Finish off the planking as described in the last chapter.

FINAL PLANKING

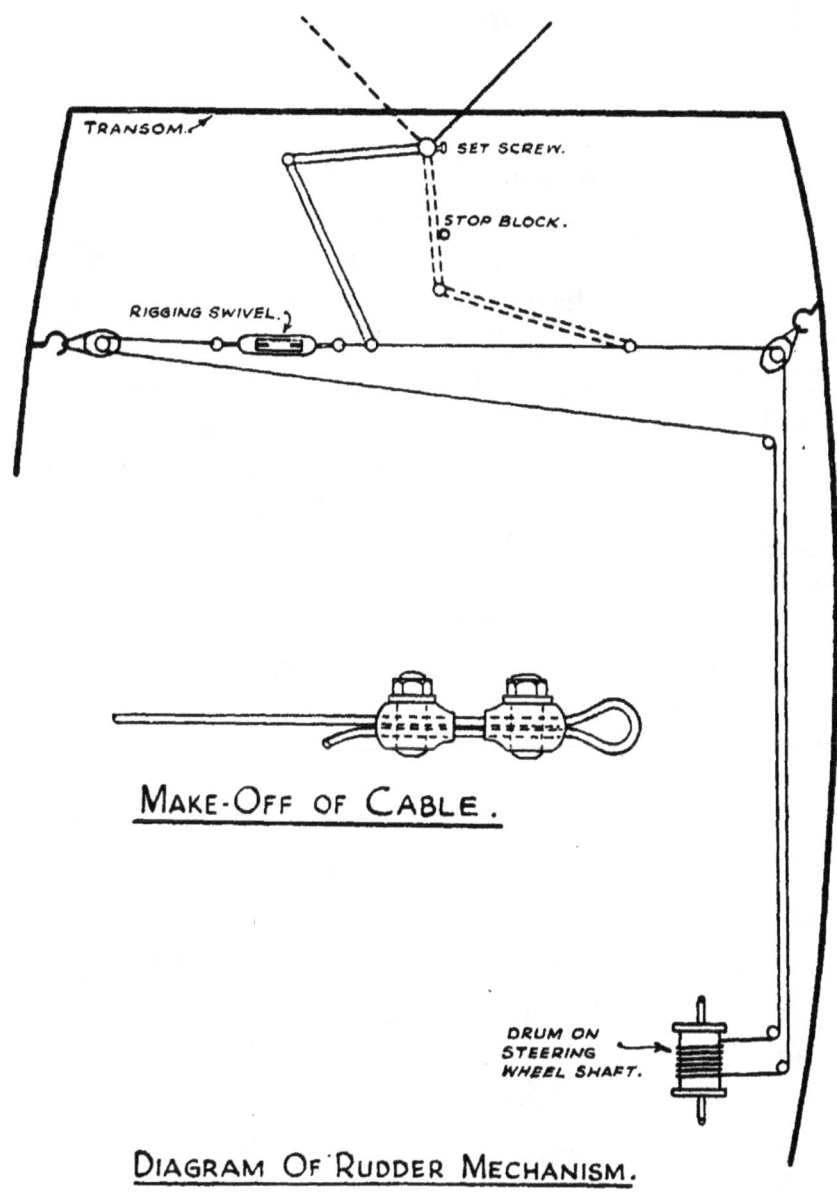

Fig. 28.—The link allows the tension of cable to be maintained.

COMPLETING THE HARD CHINE RUNABOUT

An oak rubbing strake, fitted close to the waterline along each side of the after part of the hull, as shown in Plate 10 saves much damage and paint scraping.

BULKHEADS

No doubt resin-bonded plywood is the best material for bulkheads, being always rigid and resistant to water. Cost is its only disadvantage, but it is extremely hard and if many fittings have to be attached to it one would probably wish for ordinary timber and wood screws. Take the bulkheads as far as the seam-battens, and if possible screw them to the flat side of the frames opposite to the gussets and floors. These partitions do not have to be watertight; they serve merely as compartment divisions in this runabout, mainly between the engine and the wells.

Tongued and grooved boarding is frequently used for bulkheads, and as long as the joints are upright and each board is not more than $3\frac{1}{2}$ in. wide a very sound job should result.

DECKING

There are several ways of decking a small boat, all having various advantages:

(1) Solid caulked-seam decks of varnished hardwood.
(2) Double-skin varnished decks.
(3) Tongued and grooved varnished hardwood.
(4) Tongued and grooved painted or canvased surface.

Method (1) is quite the most expensive and the most troublesome to lay, and its appearance is not so pleasing as (2). The caulking soon breaks up the varnish, and maintenance is costly. It should be of 1 in. thick hardwood planks, not more than $2\frac{1}{4}$ in. wide; a section is shown in Fig. 29 (a). A few dowels (shown dotted)

DECKING

stiffen such a deck so that walking across it will not loosen the caulking.

Method (2) is the cheapest for a varnished surface, but takes much time to lay. The first skin is of $\frac{3}{8}$-in. pine, scantily nailed to the deck beams and with rough butting seams. The top skin is of $\frac{1}{2}$-in. mahogany or teak, fastened to the beams with galvanized boat nails and to the pine skin with $\frac{5}{8}$-in. No. 5 brass screws, driven

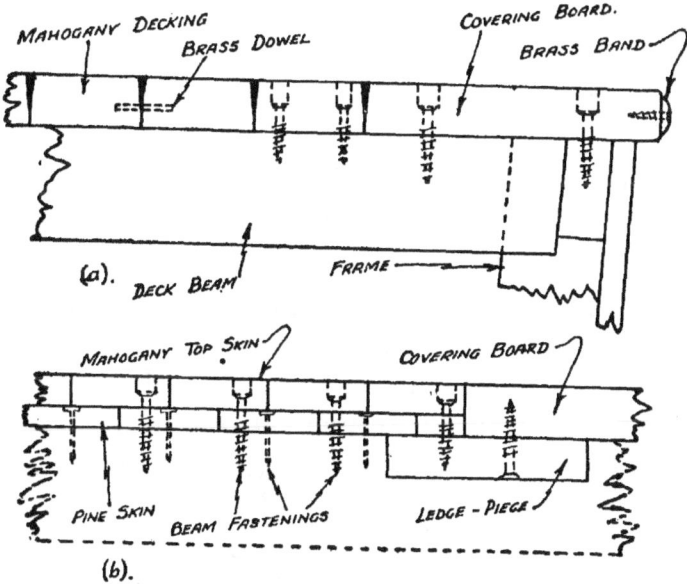

Fig. 29.—Methods of decking.

upwards between the beams. A layer of thick paint or marine glue is spread between the skins, and as long as the planks are less than 2 in. wide and all top seams kept in the centre of the pine planks there is little fear of leaks or opening seams. The method is drawn in section in Fig. 29 (b).

Tongued and grooved decks should be $\frac{3}{4}$ in. thick or, using the intermediate filler beams shown in Fig. 23, they may be $\frac{1}{2}$ in. Therefore, method (3) can be cheaper than (2) and is easier to lay, but its seams will undoubtedly

COMPLETING THE HARD CHINE RUNABOUT

open under direct sunlight unless teak is used. Secret nailing can be used with methods (3) and (4), the idea being shown in Fig. 30.

Painted or canvas-covered decks are naturally the cheapest and are quite satisfactory, but a well-built amateur hull is worthy of the little extra expense and labour which varnished mahogany or teak entail, and the appearance of the boat is greatly enhanced. For either of the last two methods use ½-in. timber and fit the filler beams shown dotted in Fig. 23.

These can be of pine or oak, sided about ¾ in. and moulded about 1¼ in. with carlines of the same scantlings. An equally good method is to make the fillers of American elm or ash, steam them and bend into position after the deck is laid. Screw-fasten the deck planks to these beams.

A covering board of some light-coloured timber like American white oak should always be used in conjunction with varnished decks, although in method (3) it may be dispensed with and substituted by the gunwale beading shown in Fig. 30. In addition to the 6-in. wide covering

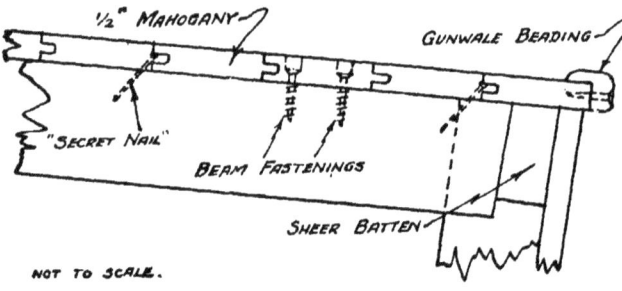

FIG. 30.—Tongued-and-grooved decking.

board, a 4-in. king plank of the same timber should be laid over the central carline, producing the pleasing effect shown in Plate 10.

Screw the covering board to the sheer batten and deck

Plate 10.—The finished boat.

COMPLETING THE HARD CHINE RUNABOUT

beams so that it overlaps the sheer strake by about $\frac{1}{2}$ in. It may be made of about three sections per side, preferably cut from curved timber. Butt joints are neater than scarphs and should be reinforced by a block of oak screwed underneath. The ledges shown in Fig. 29 (*b*) must be screwed everywhere along the covering boards and king planks to support the deck planks, which are always laid parallel to the king plank.

The hatch over the engine may have either single or double doors, which merely consist of oak frames decked over in the same manner as the fore and after decks, hinged to strips on the carlines with piano hinges. The cockpit coamings are screwed to the inboard edges of the covering board, and similar curved athwartships members cover the ends of the engine compartment hatches.

The interior fittings and equipment of a boat like this are unlimited, and as the constructional parts are pure carpentry we will not discuss them. All sorts of gadgets can be installed, and the seats and upholstery depend on the purposes for which the boat is to be used. All engine controls must be led for'ard to the steering position, which is commonly on the port side of the for'ard well.

A hand bilge pump is essential and electrical wiring should not be concealed. A windscreen is quite useless and very troublesome to fit. Its appearance may seem attractive, but omit it unless the boat is to do more than 25 knots.

PART TWO

ROUND BILGE CONSTRUCTION

CHAPTER VI

TIMBERS, TOOLS AND SOME USEFUL HINTS

BEFORE passing on to the intricacies of small round bilge construction there are a few general items of interest concerning all types of build, which should be treated. Most amateur boatmen do not know the name or use of any but the commonest timbers, and the choice of the right wood for each member is very important in boat building, just as the examination of each piece afterwards controls the success or failure of the result. Besides supplying a written order for timber, always visit the sawmill and tell the manager or foreman exactly what you want and the purpose for which it is intended. Always get the stuff planed—it costs very little extra and will save much trouble.

Where the exact dimensions of timber are important, such as in the case of planking and decking, always state the *finished* thickness required. Timber is known in the trade by its measurements in the rough state, and when planed will be approximately $\frac{3}{16}$ in. less for dimensions up to 2 in. and may be $\frac{3}{8}$ in. or more below the specified sizes for the bigger scantlings.

Timber which shows obvious signs of attack by wood worms should not be accepted, but if only a few worm holes are evident, it need not necessarily be rejected, but should be treated instantly with a generous application of paraffin, worked well into the holes.

The table on pages 85 and 86 shows the woods most used for boat building, with their various properties and uses. The recommendations are very arbitrary, as so much depends upon the size and type of the boat, and it is not always possible to obtain the best timber.

Although imported timber is generally more suitable, a small boat may quite well be made entirely of British woods. The commonest British soft wood, larch, will make fine planking for a small boat, provided it be chosen with great care and is thoroughly seasoned. Generally the larch found in the South and Midlands is inferior to the Northern and Scottish varieties, and though the Scottish boat builders claim their local variety to be " unbeatable " for planking and stringers, one rarely sees it used in the South. However, if varnished, it presents a very fine appearance, and it should not be scorned in favour of some unknown straight-grained imported soft wood.

English oak is undoubtedly one of the best timbers in the world for framing and heavy members. Properly air-seasoned for several years, it will last indefinitely under the most trying conditions, and its varnished appearance is superb. Kent and Sussex oak is claimed to be the finest, but an inferior species is difficult to find in Britain. English elm can be bought in such fine sizes, and so cheaply, that it appears ideal to inexperienced eyes. Avoid it for any member in a boat; its only possible application is for the keels and garboard strakes of big vessels which are rarely hauled out and never enter fresh water.

Lastly, a word about mahogany. This timber has an attractive appearance when varnished, and can be used for planking and framing. In Britain its cost usually limits it to ornamental work; it is often used on small boats for decks, coamings and hatches. Always use Honduras for these; Philippine (Pink) is too light in colour, although it can be darkened by staining, but either

TOOLS

the Red or Pink variety is inferior in appearance when varnished; Lagos is too dark, and is heavy and difficult to work. The main use of Lagos mahogany in boat building is for keels and dead woods. Honduras is fine timber, but to obtain a really fine surface for varnishing is extremely difficult and requires some practice. It can rarely be planed to perfection with an adjustable steel plane, mainly because the backing iron will not usually move far enough towards the edge of the blade. Therefore use a wooden smoothing plane and move the backing iron to within $\frac{1}{64}$ in. of the blade edge, as shown in Fig. 31. Use the plane very heavily. Take short curved

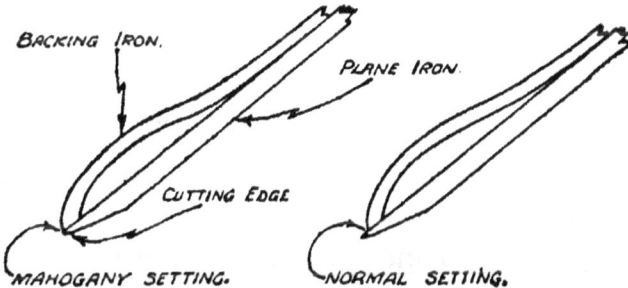

Fig. 31.—Honduras mahogany is difficult to plane.

sweeps, always bearing down on it with your fullest pressure, so that once started, a shaving always follows right through to the end of the sweep. A fine setting of the blade is essential or the surface will rip; there is no continuous " way of the grain " with Honduras mahogany.

When ordering teak make sure you get Rangoon, Moulmein or Burma teak; not Borneo or Java.

Tools

A minimum list of tools for the amateur builder is rather impracticable to compile because, although a good builder could probably manage with a saw, chisel, hammer, plane, screwdriver and a few drills, a lot more

TIMBERS, TOOLS AND SOME USEFUL HINTS

tools make things very much simpler and speed up work considerably. If the many little-used tools can be borrowed, a lot of expense is saved, and some can be used which would otherwise be classed as non-essential. The following is a list of tools which should be acquired before starting on a small boat; a few of the extra items which will prove handy are mentioned later:

Saws :
- 1 medium hand-saw with fine teeth (or 20-in. panel saw).
- 1 large tenon saw.
- 1 small frame saw and/or a compass saw.
- 1 hacksaw.

Planes :
- 1 shooting or jack plane, preferably wooden.
- 1 smoothing plane, wood or steel (Stanley No. 4).
- 1 spokeshave, wood or steel.
- 1 small block plane (Stanley No. 102).

Chisels :
- 1 wood chisel, $1\frac{1}{4}$ in. or $1\frac{1}{2}$ in. (bevel edged).
- 1 wood chisel, $\frac{1}{2}$ in.
- 1 mortise chisel, $\frac{1}{4}$ in.

Screwdrivers :
- 1 cabinet type with $\frac{1}{4}$-in. tip.
- 1 cabinet type with $\frac{3}{8}$-in. tip.

Drills :
- 1 carpenter's ratchet brace.
- 1 hand-drill to use metal-work drills.
- 3 Screw auger bits, $\frac{1}{4}$ in., $\frac{3}{8}$ in., $\frac{1}{2}$ in.
- Morse drills, all sizes from $\frac{3}{64}$ in. by 64ths to $\frac{1}{4}$ in.; also $\frac{3}{8}$ in. and $\frac{1}{2}$ in., to be used in the brace.

PROPERTIES AND USES OF TIMBERS FOR ALL TYPES OF BOATS

Species of Timber	Colour	Weight lb. per cu. ft.	Holding for screws	Planing	Life, wet	Life, dry	Approx. cost per cu. ft.
Oak, English	Brown to Straw	53	Excellent	Tricky	Very long	Very long	15/-
Oak, American White	White	42	Excellent	Easy	Long	Long	15/-
Elm, English	Dark Brown	40	Poor	Difficult	Moderate	Short	10/-
Elm, American or Canadian Rock	Straw to White	45	Fair	Smooth	Long	Long	14/-
Elm, Wych	Yellow	45	Fair	Smooth	Long	Long	15/-
Teak	Brown to Yellow	47	Good	Easy	Very long	Very long	22/-
Mahogany, Honduras	Pink to Brown	40	Poor	Tricky	Moderate	Long	18/-
Mahogany, Philippine (Red)	Pink to Brown	38	Poor	Tricky	Moderate	Moderate	17/-
Mahogany, African	Red/Brown	45	Fair	Tricky	Long	Long	20/-
Greenheart	Brown to Green	70	Excellent	Difficult	Very long	Very long	22/-
Jarrah	Dark Red	60	Good	Easy	Long	Long	20/-
Ash	Straw to White	44	Fair	Tricky	Short	Moderate	12/-
Beech	White	42	Fair	Smooth	Moderate	Short	12/-
Cedar, Red	Red to Brown	30	Poor	Smooth	Long	Moderate	13/-
Memel	White	30	Bad	Difficult	Short	Short	8/-
Pine, Pitch	Deep Red and Yellow grain	48	Fair	Sticky	Very long	Long	21/-
Pine, Columbia (Imported)	Pink/Brown	35	Poor	Easy	Very short	Moderate	8/-
Pine, Siberian Red	Straw to White	35	Fair	Smooth	Long	Long	10/-
Larch	Pink/Brown	38	Fair	Easy	Moderate	Long	8/-
Fir, Douglas (English)	Brown	30	Poor	Smooth	Short	Long	8/-
Fir, Scots	Straw to White	32	Fair	Easy	Moderate	Long	10/-
Spruce, Scandinavian	White	30	Poor	Easy	Short	Short	8/-
Deal, Red	Straw to White	35	Fair	Easy	Moderate	Long	8/-

TIMBERS, TOOLS AND SOME USEFUL HINTS

Species of Timber	Lengths, average maximum	Bending	Toughness	Warping	Cracking	Uses
Oak, English	20 ft.	Good	Very good	Frequent	Frequent	Framing, Timbers, Deadwoods, Chocks
Oak, American White	20 ft.	Fair	Good	Rare	Occasional	Covering boards, Garboard strakes
Elm, English	40 ft.	Poor	Good	Frequent	Rare	Keel or False Keel
Elm, American or Canadian Rock	30 ft.	Excellent	Good	Rare	Rare	Bent timbers, Planking
Elm, Wych	25 ft.	Good	Good	Rare	Rare	Planking, Stringers
Teak	50 ft.	Fair	Very good	Very rare	Very rare	Planking, Framing, Decks, Panelling
Mahogany, Honduras	50 ft.	Fair	Fair	Rare	Rare	Planking, Interior panelling, Coamings
Mahogany, Philippine (Red)	30 ft.	Poor	Fair	Occasional	Rare	Decks, Panelling, Coamings
Mahogany, African	40 ft.	Poor	Good	Rare	Very rare	Keel, Deadwoods
Greenheart	35 ft.	Good	Very good	Very rare	Very rare	Keel, Deadwoods, Garboard strakes
Jarrah	50 ft.	Poor	Very good	Very rare	Rare	Deadwoods, Interior panelling
Ash	20 ft.	Excellent	Very good	Rare	Rare	Small timbers, Stringers, Gunwales
Beech	30 ft.	Poor	Very good	Frequent	Very rare	Chocks
Cedar, Red	50 ft.	Fair	Poor	Rare	Occasional	Planking, Decks
Memel	40 ft.	Bad	Poor	Rare	Rare	Floor boards
Pine, Pitch	60 ft.	Good	Good	Very rare	Occasional	Planking, Decks, Shelf, Stringers
Pine, Columbia (Imported)	60 ft.	Poor	Fair	Rare	Occasional	Interior work
Pine, Siberian Red	30 ft.	Good	Fair	Rare	Occasional	Planking, Decks
Larch	60 ft.	Fair	Poor	Occasional	Occasional	Planking, Decks
Fir, Douglas (English)	60 ft.	Poor	Fair	Frequent	Occasional	Interior work
Fir, Scots	40 ft.	Fair	Fair	Frequent	Frequent	Planking, Decks
Spruce, Scandinavian	20 ft.	Fair	Poor	Frequent	Frequent	Ceiling, Floor boards
Deal, Red	20 ft.	Fair	Fair	Frequent	Frequent	Interior work

TOOLS

Miscellaneous:

> 16- or 20-oz. claw hammer; 6-oz. ball-paned hammer; mallet; 2 bradawls; pincers; marking gauge; carpenter's square; bevel; 2-ft. rule; marking spike; dividers; 2 nail punches ($\frac{1}{8}$-in. and $\frac{1}{4}$-in. tips); 8-in. smooth flat file; spirit level; plumb bob; medium oilstone; 12-in. half-round rasp; outside bevel gouge; end cutting nippers; roving punch; cabinet scraper; putty knife; drawknife; adze.

A list of clamps is given in Chapter IV; sundry articles, such as a glasspapering block (cork), and paint brushes, will be required. A few of the not-so-common tools are drawn in Fig. 32. They are all simple to use, but their maintenance is one of the most difficult jobs. If used to carpentry, get a grindstone and sharpen your own chisels and planes, but otherwise get an ironmonger or carpenter to do it for you, and learn the correct methods from him. When retipping a screwdriver make sure it is done as in Fig. 33 (*a*), not as in Fig. 33 (*b*). A drawknife should be sharpened by holding on the bench and moving the oilstone *up and down* the blade. If an adze is purchased a lipped adze is recommended. This tool is very useful for the larger, sawn-frame, boats. It should be used with great care until the necessary skill has been acquired. For fine work it should only be used across the grain of the timber.

A bench with vice, and sawing trestles are necessary, and any power tools which can be acquired will be invaluable. The vice should be either an all-wood carpenter's type or a quick release steel one with wood-lined jaws. Although the band saw and circular saw are normally electrically driven, hand-turned models are definitely great assets. An electric hand drilling machine, portable saw ("Skil-saw") and planing machine are all very handy, but would not be bought specially for the purpose. An expanding centre bit is useful for making the holes for

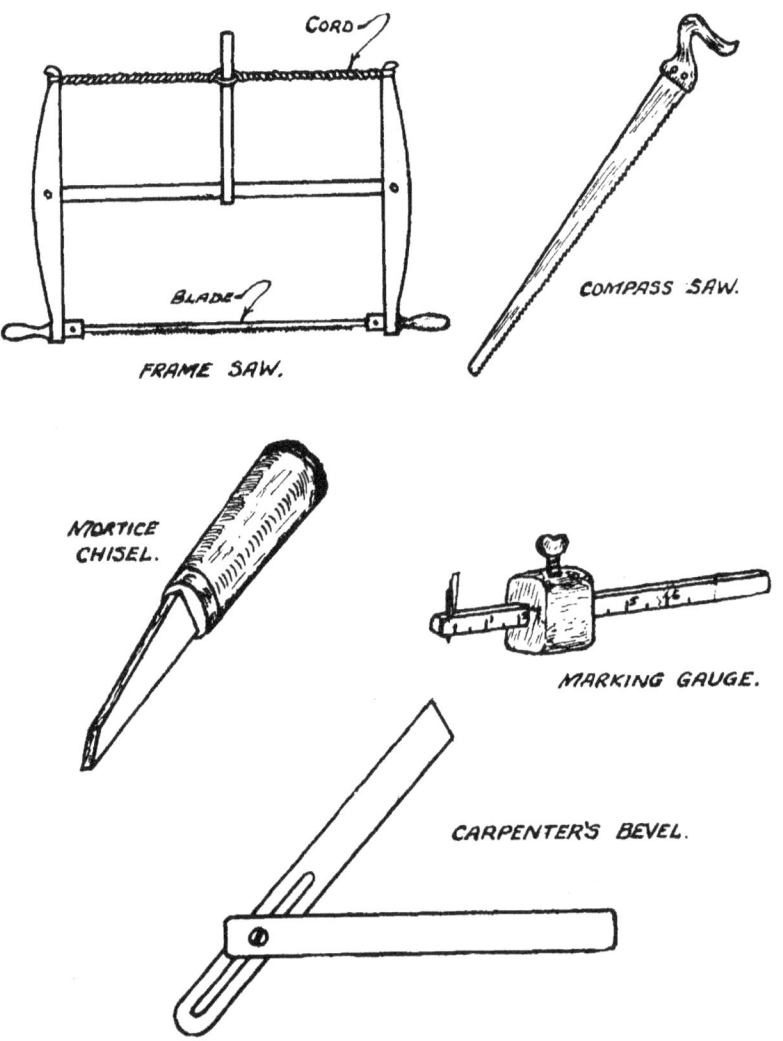

Fig. 32.—A few "not-so-common" tools.

skin connexions, but the job can be equally well done by making a ring of small holes, chiselling out and finishing with the rasp. A ratchet brace is very handy for boring

TYPES OF FASTENINGS

holes in the many awkward places found in a boat, and the larger round-shanked morse drills can be used quite well in a brace for making deep countersinks. The small morse drills are preferable to Swiss bits for boring for nails and screws. A " hold-on " is necessary for clenching and riveting—this is usually a short length of steel shafting, but a heavy hammer may be used.

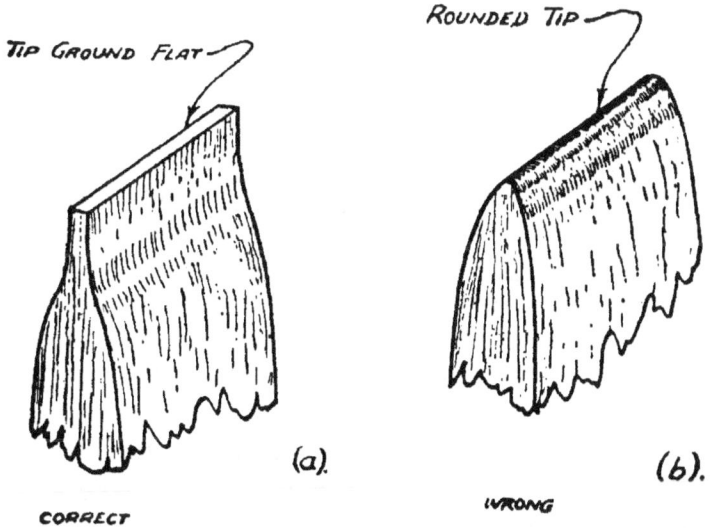

FIG. 33.—Nothing is so annoying as a badly ground screwdriver.

Types of Fastenings

So much of a boat's strength depends upon its fastenings, that a short paragraph on their choice and driving is certainly not out of place. Square copper boat nails are universally used for all purposes, but where no clenching or roving is to take place, square galvanized steel boat nails are cheaper, stronger and more difficult to withdraw, so use them for the plank frame fastenings and where nails are used in decking. Copper nails are very smooth, but unless a large enough hole is bored for them in hardwood they tend to buckle on driving. If copper boat nails are driven dead, i.e. used for other than through

TIMBERS, TOOLS AND SOME USEFUL HINTS

fastenings, they must be twisted—hold the tip in a vice, and, with a small shifting spanner under the head, give one complete revolution. The countersunk-head boat nail shown in Fig. 34 (*b*) is the type most commonly used

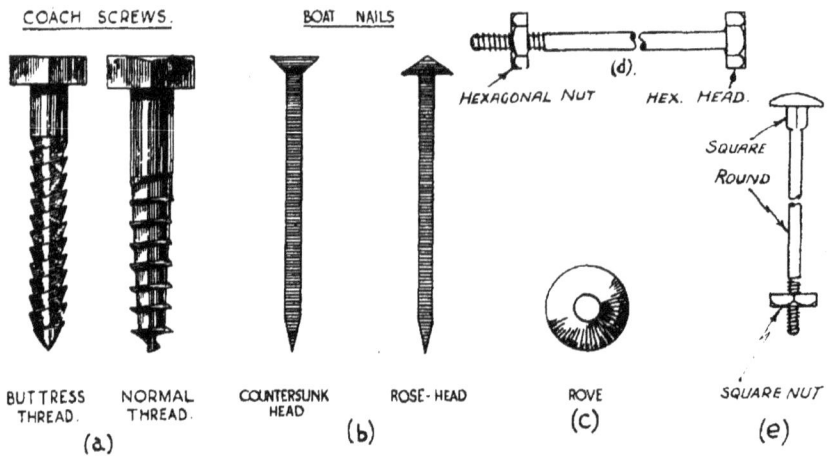

FIG. 34.—Coach screws, boat nails and bolts.

for small boats and, in some cases, is not sunk below the surface for "stopping," but left flush. A table showing the lengths and sizes of these nails and the correct roves to use is given in Appendix I.

Galvanized steel fastenings will not shear off easily, a trouble which is often experienced in old copper-fastened boats, which have "worked" through slight decrepitation.

Order boat nails by weight, specifying length, size, shape of head, and whether copper or galvanized steel. To bore for these nails in hardwood use a drill of diameter half the thickness of the nail across the corners. A large hole may be made *in the outer piece*—diameter equal to one of the sides—leaving much less nail to drive in the smaller inner hole. Moisten before driving if difficult to avoid bending. For softwood, drill half the thickness across the sides of the square nail. Bore the hole first and then countersink for the head with a morse bit just

TYPES OF FASTENINGS

the correct size. Never countersink less than $\frac{1}{8}$ in. if the hole is to be stopped later.

For varnished work, dowels may be used for stopping, in which case at least a $\frac{1}{4}$-in. countersink should be made. When driving nails into the countersinks always use a punch large enough to cover the whole of the head, or nearly so.

Screws are better than nails for all except clenched work, but to use screws everywhere else would be a waste of strength, time and money, so they are only used where a few fastenings are required to provide great strength, or where a member may have to be removed at a later date. Brass screws are always used in boats, although galvanized steel ones are satisfactory in softwood. The threads on galvanized screws are not sharp, so they are difficult to drive into hardwood, and the slot usually gets damaged and rusts. This drawback may be overcome by driving a plain steel pilot screw of the same size first, and by lubricating the final screw with oil. However, galvanized screws are very little cheaper than brass ones, so their only advantage is that they are considerably stronger than brass screws.

A few trial screws will soon teach the correct drills to use. Normally, a drill the diameter of the actual stem of the screw, about two threads from the point, is correct to bore the full length, but for screws over 2 in. in length in hardwood, two different drills are better; one, of the aforementioned diameter, to the full depth, and the other slightly larger for half the depth. A third drill, the same size as the shank (the threadless part of the screw) may be sunk to the depth of the shank. See table in Appendix I. Clear the exit ends of the holes in the planks with a chisel or large morse bit so that the seating on to the seam battens will be perfect. Driving will always be simplified if a drop of oil is put on the thread beforehand.

TIMBERS, TOOLS AND SOME USEFUL HINTS

Coach screws are mainly used in boats for securing the mechanical parts to timber. They are made in brass, steel and galvanized steel in numerous sizes and lengths, and have either square or hexagon heads so that they can be driven with a spanner. The normal kind have the usual wood-screw thread. They are also made with a buttress thread as illustrated in Fig. 34 (*a*); this type is designed to be driven part way with a hammer before twisting home. The advantage over wood screws is that they can be made a very tight fit owing to the leverage obtained with a spanner, but keep well away from the edge of wood so that there is no risk of splitting.

Bolts are very important and many types are obtainable. Galvanized mild steel carriage bolts are normally used for most boat-building purposes; they are made as in Fig. 34 (*e*) specially for use in timber. Both the heads and the nuts can stand less stress than the stem. Unfortunately the properly forged hexagonal-headed bolts, as shown in Fig. 34 (*d*), cannot often be bought ready galvanized, and to get this done to order would increase the cost considerably. Brass and copper bolts are used extensively, but the latter are very weak for their size and the former often contain flaws, so they are not ideal from the point of view of strength. The ideal shipwright's bolts are made of monel metal or stainless steel. The latter may be half the thickness of yellow metal ones and can be tightened far more; they are very easy to drive, but are rather expensive and difficult to rivet over the nuts. Tungum alloy bolts are also high tensile and good for boat building. Bore $\frac{1}{64}$ in. undersize for all bolts through hardwoods (except in the case of galvanized ones: $\frac{1}{32}$ *over* size), and $\frac{1}{32}$ in. under for softwoods. Especially in scarphs this tightness is a very important factor in the strength of the joint.

The number and size of fastenings to be used are usually

GLUES AND STOPPING

specified in the plans, for the main items. The general rule for plank fastenings to the timbers or ribs is one per 2 in. of plank width.

Glues and Stopping

Any boat can be built entirely without glue, but the use of the correct brands can increase the strength and life of a hull considerably. For all normal luting, such as between planking and seam-battens and in double-skin construction, an elastic non-hardening type of glue, such as Jeffery's " Hydro," is best. This is of a rubbery nature and brushes on very easily, especially after warming, so that a very effective waterproof seal is provided which will not break up even during movements of the timber, such as are caused by expansion and contraction. Where extra strength is required a hardening glue must be used, such as the casein, resin or phenolic types. An instance of one of these joints is the gusset or floor connection to the frames. Obviously a waterproof glue is desirable. Casein glue is practically waterproof and is usually supplied as a yellowish-white powder to be mixed with water (until it forms a gelatinous fluid) just before use. The powder will not keep and turns into a solid glassy block soon after being exposed to the air. Completely submerged in water a casein-glued joint is rather weak, but anywhere else it proves quite efficient.

Resin glues are becoming very popular nowadays and the long experimental stage seems to be over. Good brands are made by Leicester, Lovell & Co. Ltd., Ebley Mills, Stroud, and by Beetle Bond Ltd., 1 Argyll Street, London, W.1. These resin glues are completely waterproof and definitely stronger than the timber itself, provided perfect surfaces and effective clamping are ensured. Such are essential with any glue, and the smaller the space between the surfaces the stronger will the resultant

TIMBERS, TOOLS AND SOME USEFUL HINTS

joint prove. Clamp tightly and evenly, and always be very quick before the glue has a chance to commence setting.

The following recipe can be used to make a universal marine glue which is waterproof and semi-hardening and which would be suitable for joints in a small boat: Dissolve $2\frac{1}{2}$ oz. of fine rubber shreds in $1\frac{1}{2}$ oz. of coaltar naphtha by heating and stirring, then add $2\frac{1}{2}$ oz. of shellac powder. This solidifies on cooling and the pot must be surrounded by boiling water before use.

A good putty for stopping the many nail and screw holes in painted planking was mentioned on page 56, as the same putty is used for paying. For varnished work, such as mahogany topsides and decks, the stopping must be coloured to match the timber, and this colour can only be accurately obtained by trying samples to see how it appears after the surface has been filled, oiled and varnished. Therefore, for mahogany the stopping must be made very much darker than the raw wood. The addition of Vandyke brown, Bismarck brown and ochre powders will enable the correct shade to be obtained. More linseed oil will have to be added to maintain the consistency of the putty. Alternatively a stopping made with these powders and gold-size alone is quite effective and hardens very much quicker than putty.

Plastic wood is made in many different colours, and if the correct shade can be found this is the simplest of all stoppings to use. It shrinks a little on setting, so hump its surface. Always buy small tins—big ones cannot be used quickly enough, but if the stuff does harden slightly it can be softened by the addition of a few drops of amyl acetate. A stopping which is exactly the same as plastic wood can be made by mixing very fine sawdust from the wood concerned with " Durofix " liquid glue. The fine wood dust may be obtained by rubbing down a piece of

PAINTING AND VARNISHING

wood on a "sandpaper block," which consists of a sheet of glasspaper glued to thick plywood.

High-class builders often use dowels of the same timber to plug the holes in varnished work. This is a tedious job requiring skill, but it is well within the reach of an amateur. If the dowels are made in the usual long strips, with the grain running parallel to their lengths, they will always show up as darker than the surrounding plank, because the end grain absorbs the oil and varnish, so the only effective way is to make the dowels with the grain at right-angles: take pieces of the timber to any boat-yard where they have a tool for cutting these dowels. Dip the base of each dowel in varnish and tap into place with grain the correct way. To remove a faulty dowel, drill a hole in the centre and drive in a screw. If stopping is used, apply it carefully. Do not leave any on the surface of the planking, and make the stopping go right into the hole with a humped surface to allow for contraction and the later rubbing down. Make sure each hole has been thoroughly painted inside before stopping, in a similar manner to the seams.

PAINTING AND VARNISHING

Some details of painting have already been given and there is little more to say about the application of paint, but before even priming is put on, any knots in the timber should be treated. This is done by "killing" them with a coating of slaked lime and painting the places with "knotting," which can be any quick-drying varnish, shellac dissolved in methylated spirits, or a mixture of red lead and gold-size. Then paint over these with the priming.

Ordinary boat paint can be bought ready mixed or made by adding various colours to a standard white lead paint of reputable brand. Always use colours " ground in oil," not the powders, and adjust the consistency of the resultant mixture by adding raw linseed oil and a little turpentine, but no driers. Always keep paint well stirred while using it. You can mix your own paint entirely by starting with a base of white lead (ground in oil) and dissolving this in linseed oil and turps. Add a small quantity of driers (proportion arbitrary), filter through several layers of muslin and you can add your colours as desired. For a white finishing paint always use zinc white in place of white lead, as it does not discolour, and always add a few drops of ultramarine in oil, or blue paint, to brighten the white. To mix a glossy paint reduce the amount of linseed oil by one-quarter, and add an equivalent amount of clear copal varnish and more driers.

Enamel is difficult to make properly. White enamel is mixed with poppy oil (transparent) instead of linseed, and about 60 per cent. of clear copal varnish. Commercially, it is very thoroughly ground and mixed in a machine, and filtered many times. The best enamels are made with a special varnish which is unobtainable in shops.

With patience, rubbing down each coat carefully with fine glasspaper and brushing out each coat very thoroughly and evenly, there is no reason why the amateur should not obtain a professional finish.

Varnishing is simple, provided all the rules are observed. First, the surface must be made dead even and silky, by glasspapering and/or scraping. Secondly, the grain of the wood must be "filled." This may be done with one of the special coloured paste fillers—which is applied on a cloth pad and, after well rubbing into the grain, is cleaned off with cloths. If, however, there are contrasting colours of wood such as oak and mahogany, a good plan is to use a plaster filler, such as "Alabastine." This is a white powder which is mixed to a thick paste with water and applied in a similar manner. When set,

PAINTING AND VARNISHING

scrape off or rub down with glasspaper. The application of raw linseed oil renders the filler transparent and darkens the wood. An interval of 24 hours should then be allowed before varnishing.

Thirdly, make sure the air is never cooler than about 45 deg. Fahrenheit, and that there is no dust in the workshop. If necessary, sprinkle the floor with water. Fourthly, apply the varnish by cross brushing so that a very thin film covers the whole surface evenly, and always finish each section by brushing in the direction of the grain of the wood. Fifthly, wait until the varnish is hard (about 5 days for copal varnish) and rub down each coat, except the last, with fine glasspaper (preferably " wet " glasspaper, used with water). Apply two coats for interior and at least three coats for exterior work.

Use only the finest copal body or yacht varnish for boat work ; it takes a long time to harden, but is worth waiting for. Choose the warmest and least humid weather possible : if too cold, the surface will possess a permanent " bloom," or may never harden completely. Always store varnish in the house during the winter ; once frozen it will never harden.

CHAPTER VII

BUILDING A 3-TONNER—(1)

AS previously stated, it is impossible to say whether chine hulls are easier to build than round bilge. Personal opinion is the deciding factor, so having read about hard chine build, the following text should enable the prospective builder to decide the type of boat most suited to his ability.

Many different ways of building small round bilge hulls are possible. The timbers may be—

(a) Steamed and bent before planking.
(b) Steamed and bent into position after planking.
   NOTE: Method (a) is more usual, except for clinker, and in either case the ribs may be a single thickness or two or more laminations.
(c) Widely spaced sawn frames, with one, two, or more bent timbers between them.
(d) Angle iron, bent hot, with wood keel and planking (composite construction).
(e) All sawn from oak crooks.

The planking may be—

Clinker, carvel, or one of the multi-skin modes of construction, i.e.: Double diagonal; diagonal and fore and aft; " Ashcroft "; double diagonal or Ashcroft *and* fore and aft (three skins).

NOTE.—In all multi-skin construction stringers may be used without timbers, or the planking may be laid over the usual framing. (If all the framing were removed from a double skin vessel it would still retain its approximate shape). The type of planking to be chosen depends mainly upon the size and type of boat.

DINGHIES

No mention has been made of all-metal construction which, in steel and various alloys, is becoming increasingly popular. In most respects it is more difficult to tackle than wood, and is not recommended for the first attempt of an amateur.

Resin-bonded plywood is only of interest to the amateur for the planking of small hard chine hulls, and perhaps for decking and internal decoration. This material would only prove economical for round bilge boats if being mass produced.

Dinghies

Although all the practical details of round bilge construction will be treated later in this chapter when the 3-ton 17–20-ft. sloop is considered, the method of building dinghies is rather different, and the following notes will indicate where most of these differences lie.

Dinghies, in common with larger craft, can either be built right way up or upside down; the following description is of upside down type construction. The moulds and transom are erected on twin stocks (say 6 in. by 2 in. on edge, say 2 ft. 6 in. apart), keel notches uppermost, and the stern and keel members are fitted into their notches. The clinker planking is all fitted and lightly nailed to the moulds where necessary, all land nails being driven, and riveted up as required. The hull is then turned over and the ribs are steamed and bent into their previously marked positions—the planking nail holes having been already drilled, and the nails driven both ways from the keel up while the hot timber is being manipulated (the timber need not be drilled for the nails if they are driven within a short time of leaving the steaming box). Any floors are fitted, and after the removal of the moulds the gunwales are steamed and placed, and the risings and thwarts fitted. NOTE: Stringers are

BUILDING A 3-TONNER

seldom used in a clinker boat, but risings, which are similar to stringers at thwart level, carry the thwarts.

A typical midship section of the keel is shown in

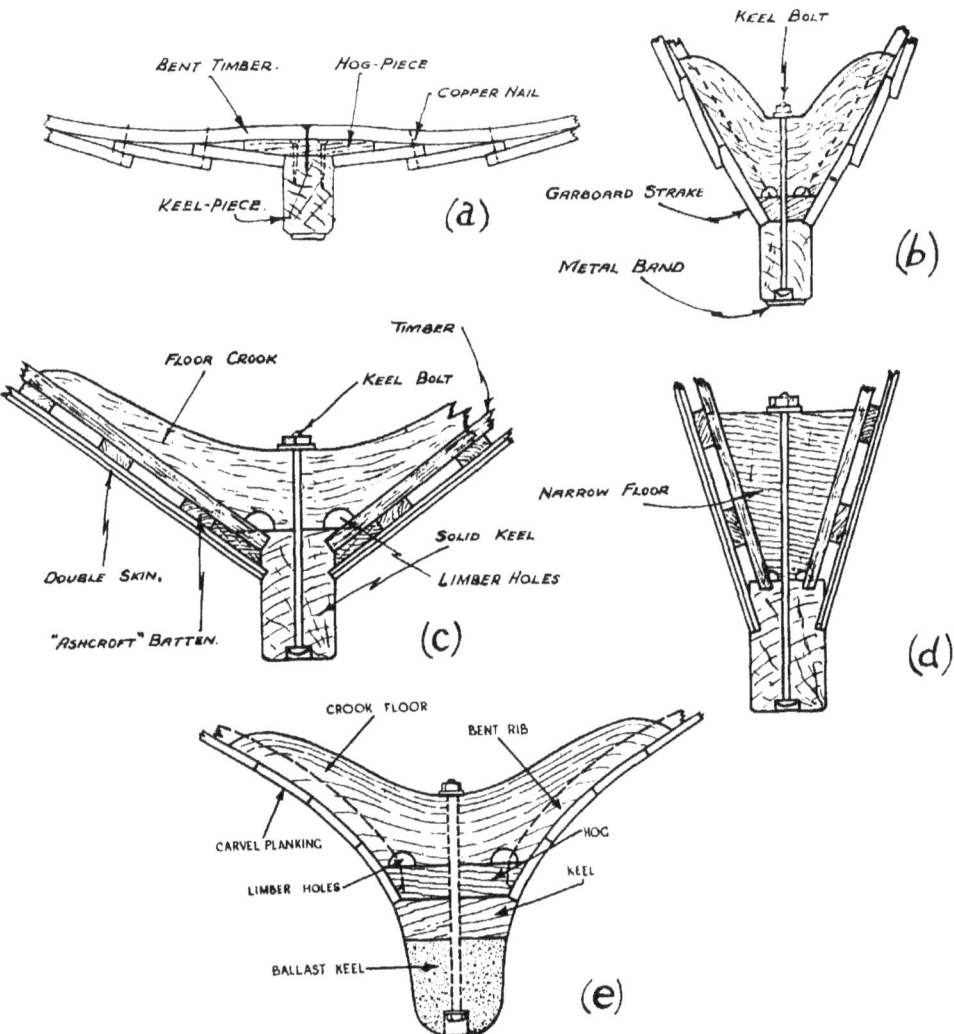

FIG. 35.—Five different types of keel construction.

Fig. 35 (*a*), and a for'ard section, with oak floor in Fig. 35 (*b*). The long steamed timbers can be made in two half-lengths, overlapping on the hog-piece, to which their heels must be firmly nailed or screwed. A keelson member can be fastened on top of the timbers, directly above the hog-piece, but without filler blocks between the ribs a nasty dirt-collecting gap is formed, so a keelson is only fitted to a boat requiring great strength in her keel, such as one that frequently has to land in the surf on a pebble beach. Such an extra strong boat often has her timbers joggled to fit accurately to the irregular inner surface of the planking. Floors are not essential in a dinghy, but in any larger boat they are very advantageous, giving strength and seam-tightness. Strong floors are necessitated in the way of an inboard motor. With clinker planking, the floors are joggled to fit snugly, and they are always fitted between a pair of ribs, not on top of one.

The floors should be through-bolted to the keel and the planking fastened to them by at least one screw in each strake, with single bolts in lieu of screws at the extreme tips of the "wings." The hog-piece is fastened to the keel with stout screws placed zig-zag; each bent timber is usually nailed to the hog with one G.I. nail driven whilst the timber is very hot and then to the lands (or laps) of the planks by copper nails right through from outside as previously indicated: these fastenings are usually clenched.

Small sailing boats, with reverse curve in the garboards, or with V-shaped bottom sections, are sometimes built with a solid keel, as in Figs. 35 (c) and (d). Planking on the Ashcroft system is shown in these drawings, and carvel in Fig. 35 (e), where a ballast keel is incorporated. Carvel planking can be used for dinghies, without seam battens, but is usually heavier, and more costly in upkeep.

In the Ashcroft system, the diagonal skins (inner and outer) run parallel with one another, each outer one covering the joint between two inner ones.

BUILDING A 3-TONNER

The " double-diagonal " method is much stronger than " Ashcroft " but has the one great drawback in that it is very difficult to repair. It means extensive stripping to patch up a hole and the result can never be as strong as the original construction. If the hull happens to rot along the keel, or along the waterline, dozens of planks may have to be renewed, whereas with any type of fore-and-aft planking, renewal is a simple matter and involves no loss of strength.

Clinker planking is marked out much as already described for carvel. The lapped edges are bevelled to a width of about one and a half times the thickness of the planking. The land nails are riveted on roves (either one or two between timbers), and the nails attaching the planks to the timbers are either riveted on roves or stapled with the clenched ends of the copper nails, on the inside, at an angle of about 10 or 15 deg. with the run of the timber. A further note on clinker planking is given in the next chapter.

PROCEDURE IN BUILDING

The *Yachting Monthly* 17–20 ft. Sloop [1] is an ideal design for an amateur: it will provide a good example for building details.

The building is simplified by a complete lack of reverse curves in the sections, leaving a large deadwood area which is not planked over. The coach roof extends right to the topsides, and although this means a little difficulty with the moulds in that vicinity, much material and labour are saved.

[1] A full description of this craft appeared in the *Yachting Monthly* for September and October 1946, and Fig. 36 shows only some of the construction details. Complete sets of plans for amateur building can be obtained from The Editor, *Yachting Monthly*, 3 and 4, Clements Inn, London, W.C.2, price £4 4s., including the designer's royalty.

PROCEDURE IN BUILDING

A complete lofting of the lines is indicated because the profile is needed for the stem and deadwood members, and the half-breadth waterlines, buttocks and diagonals should be checked against the body plan curves to enable accurate moulds to be made.

Start on the main framing first, which consists of the stem, keel, hog-piece (or keel batten), sternpost, deadwood and stern knee, a similar partly finished assembly appearing as in Plate 11. Some notes on the choosing of timber for these members are given in Chapter IX. English oak is the best wood for all framing, being reasonably priced and procurable, as well as extremely tough and durable, if well seasoned.

Well-seasoned oak is scarce these days, and the use of " dry " but not *well*-seasoned timbers in the fore and aft main frame members is not a very serious fault as, in seasoning in position, they shrink back upon themselves. Planking and decking are the parts which *must* be really well seasoned.

Natural crooks (or sweeps) must be used for the stem, which is sided $3\frac{1}{2}$ in. for most of its length, but widens out to 10 in. where it meets the keel. In order to get the scarph accurate, a light wood pattern or template had best be made for the stem. All the other main framing members are straight, and do not need patterns. Having cut the scarph faces, lute, clamp them, and bore the four bolt holes, but only fit three of the bolts, because the fourth will pass later on through the stringer breasthook. When boring bolt holes, always choose the correct drill, screw auger bit, or auger, by making trial holes in some scrap of the same timber, into which the bolts are a tight driving fit. Make the stem notches for the hog-piece and keel, and then chisel out the plank rabbet and chamfer the cutwater. An adze is the best tool for the latter operation, and as it can be used profitably for many boat

Plate 11.—A keel, stem and sternpost assembly.

PROCEDURE IN BUILDING

building jobs, the amateur should acquire the technique at an early stage, by trimming the sapwood off oak crooks, or by practising on scrap timber. If possible, learn a few wrinkles from a shipwright. The chamfer is usually finished off with a plane at a later stage, when the hull is being cleaned off.

It is wise to allow at least an eighth of an inch extra timber in the back of the rabbet and on either side of the cutwater lines—the former to be removed, as necessary, during planking, and the latter after planking is complete. The stem should be repeatedly oiled with linseed oil or painted, two coats (except the end grain at both ends) to help it to preserve its shape. It should then be supported in a sheltered and shady position until the deadwood and sternpost have been bolted to the keel.

Note that the keel increases in its moulded width in the way of the centre-board, but where the deadwood comes it is a constant 7 in., the same width as the deadwood and sternpost. The system of bolting all these pieces together is shown on the drawings, but it is quite arbitrary, and any number of scarphs may be used in the deadwood members.

To cut the centre-board slot, bolt or clamp the pieces concerned together, and after accurately marking the position of the slot on the top of the hog and the bottom of the keel bore a succession of $\frac{5}{8}$-in. or $\frac{3}{4}$-in. holes with a brace and bit from the hog down to a little more than half the thickness, then, after turning the assembly over, from the keel down to connect up (approximately), taking care to angle all holes slightly inwards. Bump out the core with a hammer and finish off the slot, with a wide paring chisel, checking for " square off " by slipping a try-square through it at the ends and centre. An alternative method favoured by some builders is to cut the various slots for the centre-board separately in each piece

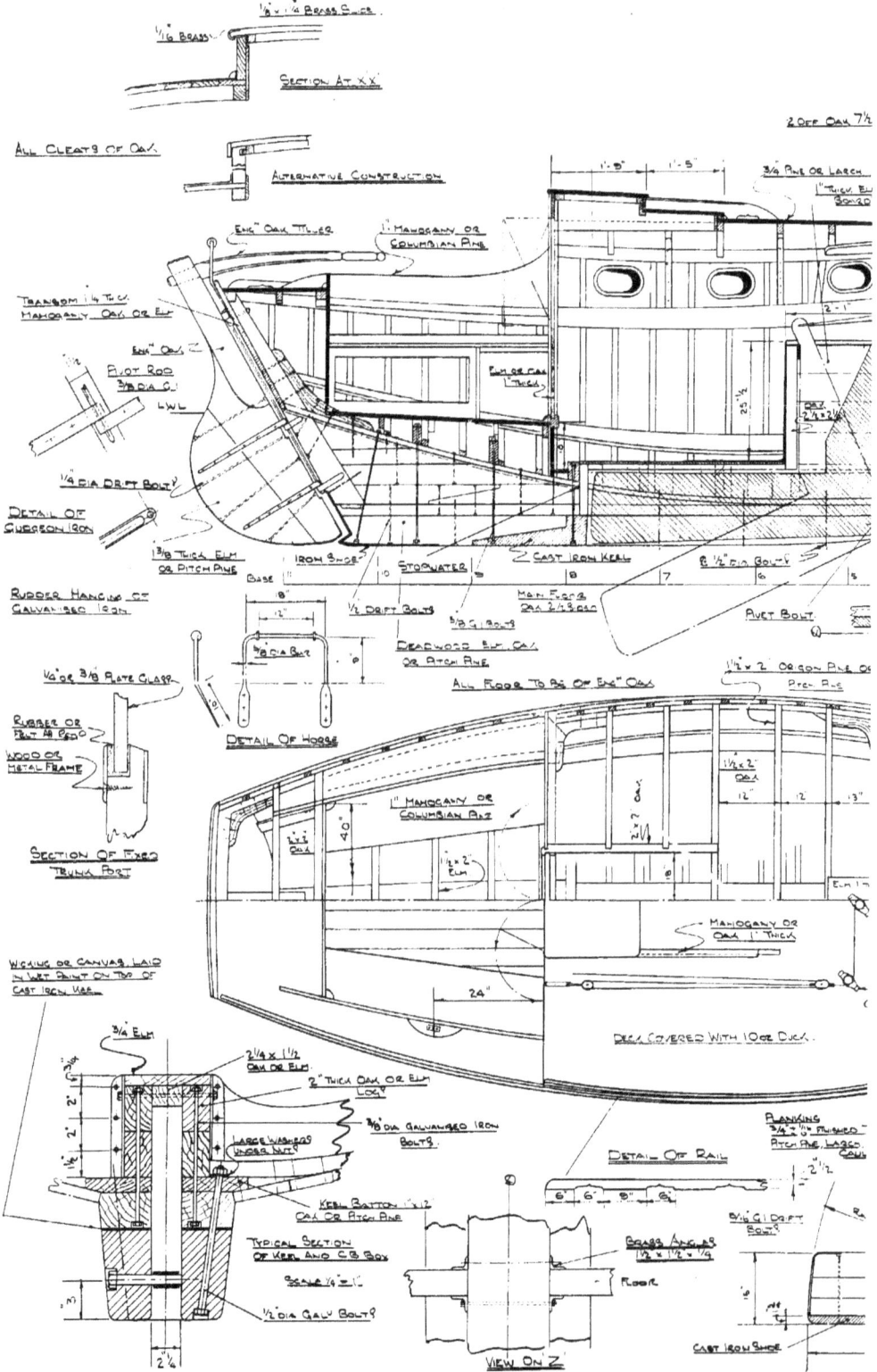

FIG. 36.—Some construction details

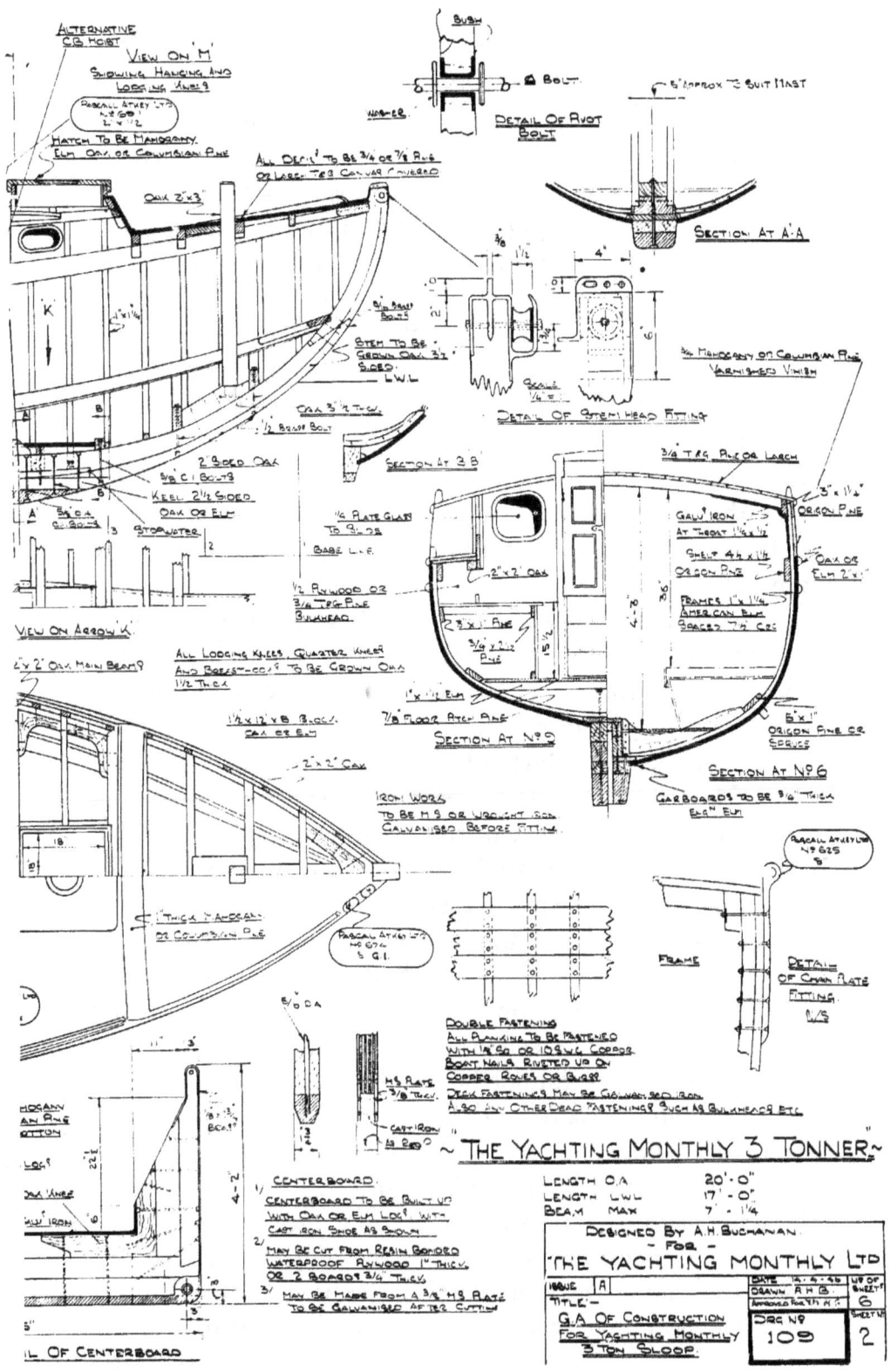

the *Yachting Monthly* 3-tonner.

concerned by boring a row of 2¼-in. holes and chiselling between.

Chamfer off the edges of the top of the deadwood and the underside of the hog-piece to form the garboard plank rabbet. All the bolts, except those attaching the false keel and the floors, can be driven. A shoulder is to be made on the sternpost, where the hog-piece forks round it, and the sternpost then continues up the transom at this reduced width, with its heel stump-morticed adequately into the keel. The hog-piece may now be fitted, with the stern-knee bolted on top of it, as shown in Fig. 38, and the stem fitted with temporary bolts. Permanent bolts cannot be used until the ballast keel and floors are fitted.

Next fit those parts of the centre-board case which abut to the keel, and bolt them through to the keel. All the members mentioned so far must be thoroughly treated with some wood preservative, such as creosote, Solignum, or Cuprinol, with particular attention to seams. This centre-board case must be perfectly watertight, meaning a perfect fit to the curve of the hog-piece, and a thorough luting of all seams with liquid marine glue, Stockholm tar, or white lead paste. Keep the bolt spacings no more than 9 in., and fit the angle brackets to which the floors beside the centre-board case are bolted. These jobs completed, bore for the stopwaters (one of which should be fitted within the chiselled planking rabbet wherever a scarph crosses the rabbet below the waterline) and drive in ½ in. diameter softwood dowels, well luted with paint or marine glue.

A separate lofting is required for the transom, because it is sloping in relation to the station lines. To do this, draw the transom centre-line the correct length, as measured along the sternpost, divide it into an arbitrary number of equal spaces, and draw lines through the

PROCEDURE IN BUILDING

points, at right angles to the centre-line. Divide the lofted transom centre-line into an equal number of spaces, with similar athwartships lines, and transfer the distances from the centre-line to the edge of the transom to the corresponding lines on the new pattern, obtaining a true outline when all the points are joined by a curve. An exaggerated sketch of the process is shown in Fig. 37.

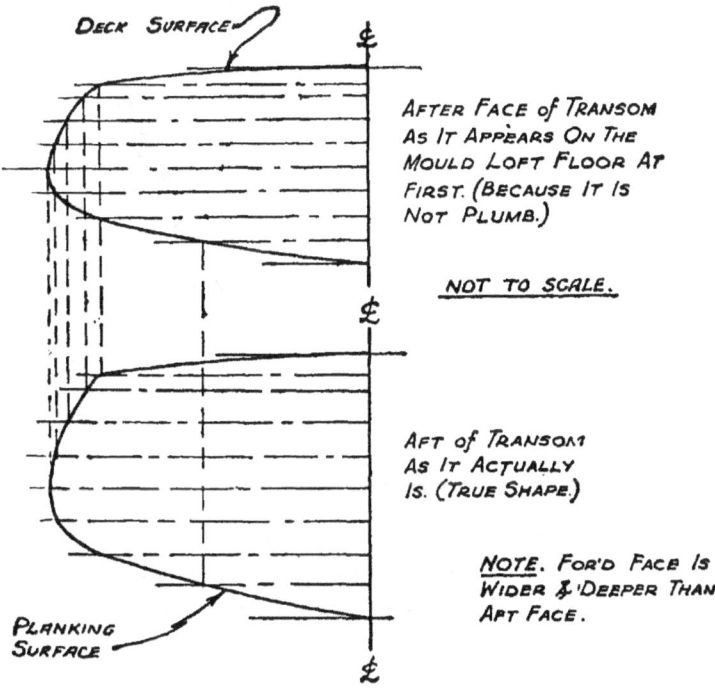

Fig. 37.—Re-lofting a raked transom.

Oak or elm should be avoided for the transom. Mahogany, teak or pitch pine are best. Build it up with planks not more than 8 in. in width, making caulked seams, and fit a frame all round the edges, from the deck beam to the sternpost, ¾ in. from the edge of the transom, and of 1½-in. oak about 3 in. wide. This frame takes the hood-end plank fastenings and the ends of the stringers. Note that the transom itself is lofted to the outside of the planking, so that the plank ends seat into the rebate shown in Fig. 38. Alternatively, the frame may be finished flush with the transom planking, so that the hood ends are left bare and offer a caulked seam round the transom.

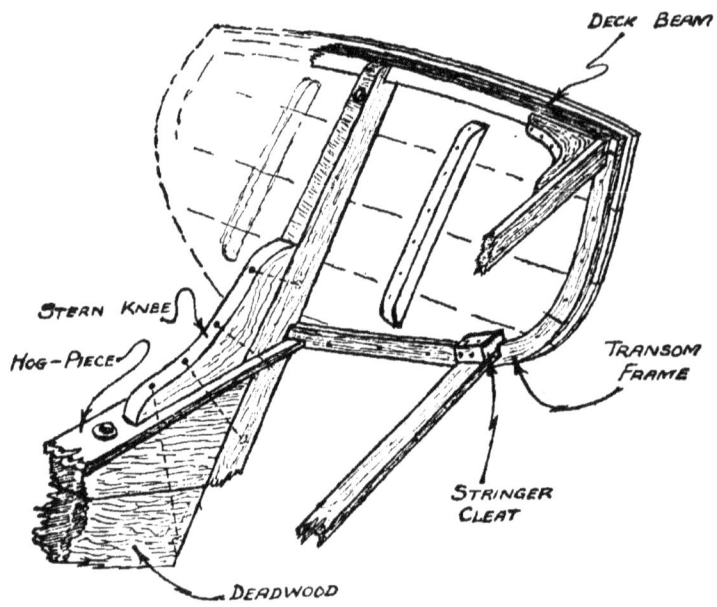

FIG. 38.—The stern assembly.

NOTE.—The lofted outline of the transom will be to the edge of the outside of the planking, and the bevels must be applied to get the larger inner edge of the transom planking.

THE BALLAST KEEL

Before finally bolting the transom through to the stern-post, get the ballast keel cast and fitted, as this operation will be easier when those parts adjacent to the casting can be lifted on to it alone. The casting is quite a complex one varying in width and depth, and having a slot for the centre-board. Suitable foundries will make their own pattern from the drawings, but the amateur may make his own pattern, of light wood, the correct shape of the casting required, but increasing all measurements by $\frac{1}{10}$ inch per foot, to allow for shrinkage of the casting. Use thick enough timber to get the rounded corners, though the pattern can be built of solid wood, in several laminations, perhaps more easily than by using a thick framework covered with thin planking. Mark the bolt hole positions very accurately, and glue studs of wood of the required diameter at the top and bottom of each position, projecting about half an inch. Make a wood core to show the diameters of the holes and the positions of the shoulders for the heads of the bolts. The hole below the shoulder may be square or round, according to the type of bolt, and a separate core will be needed for the centre-board pivot-bolt hole. Glasspaper the pattern smooth and give it a couple of coats of shellac varnish before handing over to the foundry. If possible, get the casting sand-blasted before delivery or, failing that, get it pickled, or cleaned with a power-driven wire brush. Then apply a few coats of bituminous or red oxide paint, including the insides of the holes. Clamp the casting to the keel on its packing of canvas soaked with thick white-lead paint, fit the after false keel to it, bore all the keel bolt holes, and fit all those bolts which do not connect to floors.

Building Site and Erection

All the work so far could have been executed in a backyard, and the moulds can be made in the same place. In fact, were it possible to lift the completed hull on to a truck, the entire job could be accomplished any distance from the water. Making as much as possible at one's home, and then erecting on an improvised slipway is a most convenient method, but if the loft floor is in one place it may later be wanted in the other, so a single building site has advantages. It all depends on the distance from one's house to the building yard. The ideal

BUILDING A 3-TONNER

site is adjacent to a boat-yard, from which tools and strutting timber may be borrowed. Boats of this size are frequently built in a shed. This eliminates the weather problem, and simplifies strutting, but launching difficulties are apparent unless the shed is a suitably designed one.

We will presume erection is to take place on a river bank, and that the hull is to be launched stern first. The keel must be blocked up to enable the floor bolts to be pushed up from below the keel, and the slip must have a declivity to enable the hull to slide into the water. This declivity is best made about 1 in 8. Temporary blocks can be placed under the keel to get it erected, but a better way is to fit up the actual slipway (consisting of the standing ways upon which rest the sliding ways and cradle), at the start, and erect the hull on its launching cradle. The slipway need not extend right down to the water until just prior to launching; make it merely the length of the boat. Set out the required slope for the slipway with a mason's level and a staff marked off in feet, and lay timber bearers on the ground below each slip, their size and spacing depending on the nature of the soil. Make each slip 20 ft. long and 4 ft. or 4 ft. 6 in. from its partner. For good clay soil, use 4-in. by 8-in. bearers, 3 ft. long and 4 ft. apart, but if sound ex-railway sleepers are used, they could be laid right across, advantageously linking the ways together. Dig them in if necessary, to get their tops in line, and then fasten a 2-in. by 8-in. pine plank along them with two 6-in. nails at each bearer. The effect so far is sketched in Fig. 39.

The uppermost surfaces of the standing ways should be planed smooth, as similar planks are to be placed on top of them (attached to the cradle), and the friction between the two planks should be as small as possible. The cradle is shown in Fig. 40, and four 6-in. by 6-in. beams will suffice, spaced to dodge the keel bolt holes

BUILDING SITE AND ERECTION

and to enable the centreboard to be introduced from inside, at a later stage. Lash the cradle to prevent it from sliding prematurely.

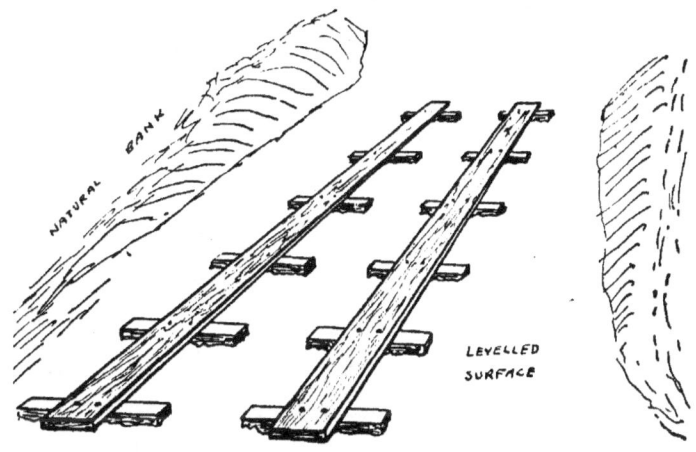

FIG. 39.—Temporary slipway.

Now, we want the waterline of our hull to be horizontal, so that the moulds can be set up plumb. Therefore, the after end of the keel will have to be blocked up some 4 ft. above the cradle, the method of blocking being shown in Fig. 40. By this means the rockered keel can

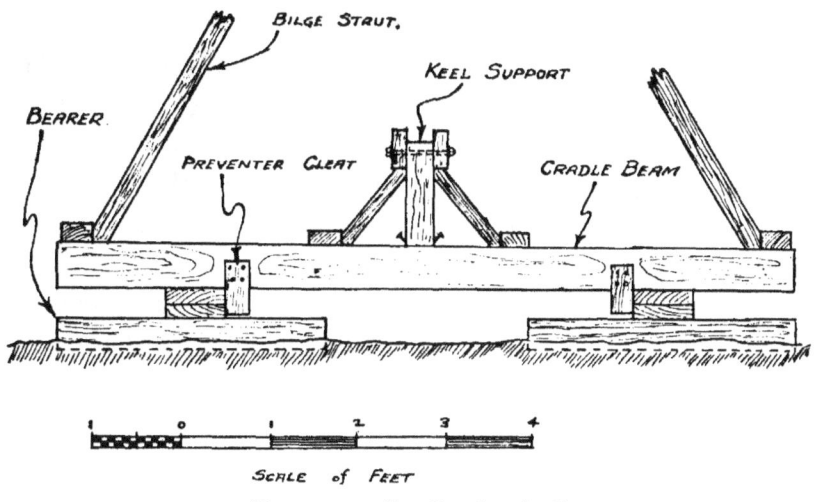

FIG. 40.—Cradle for hull.

be evenly supported, the shape being more awkward in the deep-draft version of the " *Yachting Monthly* 3-Tonner." Struts tacked to the stem and sternpost will now ensure the rigidity of the framing on its stocks.

A line must be tightly strung fore and aft under the cradle and boat at the centre line for setting-up purposes ; a wire set up tightly with a turnscrew is better still.

MAKING THE MOULDS

With the exception of Stations 1 and 3, all the moulds can be made on the body plan sections, but although bent timbers occur at these two stations, moulds could be put there, and the frames placed after their removal. It will be better to strike in two new sections a few inches away and make the moulds to them. A professional builder would use fewer moulds than this, but the amateur will find fitting frames and planking easier with a few extra moulds. The process of lifting a pattern from the loft-floor has already been described, and the moulds may be treated as large patterns. Do not forget to subtract the planking thickness from the loft-floor sections. Particularly forward and, to some extent aft, more than the thickness of the planking must be deducted, as the mould plane cuts through the planking at other than a right angle. The station lines are to be marked along the hog-piece, and the moulds abaft Station No. 6 will be set up with their for'ard sides on all the station lines, while those for'ard of Station No. 6 have their after sides on the lines, to enable the bevel to be applied similarly to the frames of the chine hull. Alternatively, the moulds can be used with no bevel at all, and set with their faces on the station lines in reverse order to the above rule. This idea saves a lot of work, but the edges of the moulds may then be crushed by the ribbands unless wedges are carefully placed to distribute the

MAKING THE MOULDS

pressure across the whole width of the edges of the moulds.

A typical mould is shown in Fig. 41. Use 1⅛-in. pine, and make it wide so that only the one tie or spalling piece is needed. A strut can be fitted between the centre

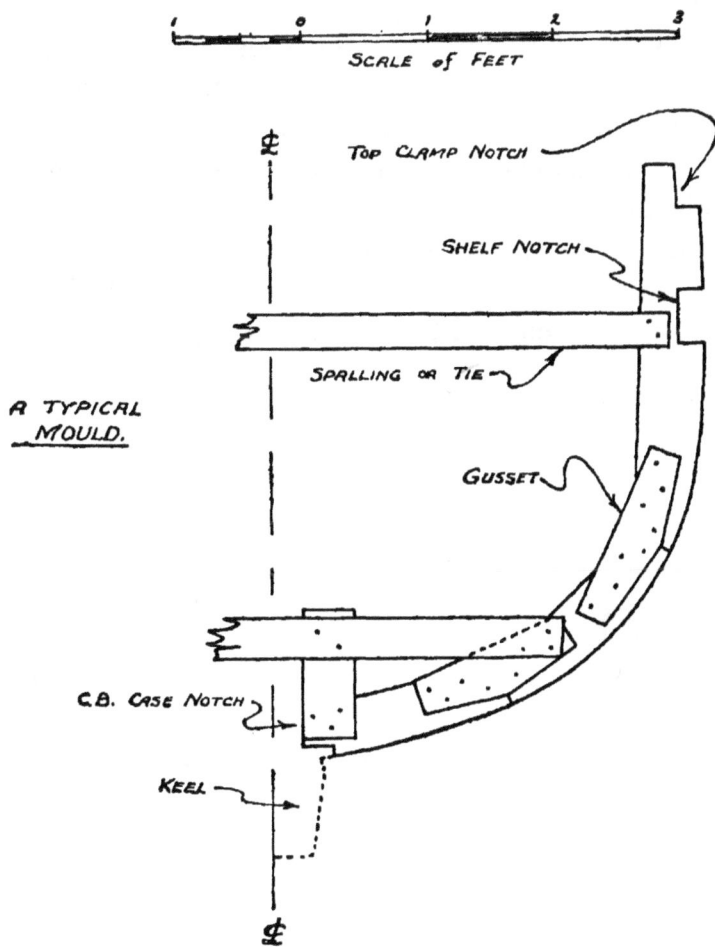

FIG. 41.—A typical mould.

of the tie and the base of the mould, but is more of an obstruction than it is worth in so small a hull. Use wood screws in all the gussets and ties, to simplify the dismantling of the moulds after planking. The big keel notch is only required in the way of the centreboard case. Elsewhere cleat the mould to the keel batten with wood blocks or metal angle brackets. Mark the centre of each tie, and after erection of the moulds, run a cord along these to get them all centralized between the stem and sternpost. Setting up each mould with a plumb-line, tack two battens along the tops of the ties and fit side struts to every other mould. This will keep them rigid enough until the shelves and side ribbands are fitted. Moulds come in for a lot of strain and must be strongly made; slats should be as long as possible, and adjacent screws or bolts placed so that they are never in the same line of grain as one another. Thickness of timber may be reduced to 1 in. or even $\frac{7}{8}$ in. *if well made.*

CHAPTER VIII
BUILDING A 3-TONNER—(2)

TO assist in the erection of the moulds and in all subsequent building operations, a system of scaffolding with gangplanks running round the hull at various heights is necessary. Make six or eight frames as sketched in Fig. 42, half of them about 5 ft. tall and half about 3 ft. These can be shifted to any position, like painters' trestles, and planks laid between them at various levels. The bases should be blocked to keep the frames vertical.

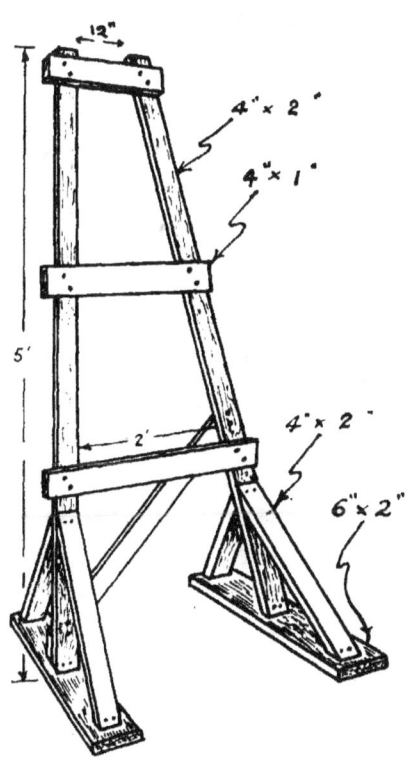

Fig. 42.—Scaffolding frames.

Shelves and Ribbands

The order of construction of a dinghy is not possible for a boat of this size, but we can use any of the following methods:

(a) Steam the timbers and bend them to the side ribbands before planking.

(b) Same as for (a), except that the shelves are put in first, and the stringers after planking and removing the moulds.

(c) Steam the frames and bend them round specially prepared jigs (usually a series of blocks screwed

BUILDING A 3-TONNER

  to the floor), trimming them after setting, and fitting to the ribbands like solid sawn frames.
- (d) Fit shelves, stringers, and ribbands into deep notches in the moulds, and then bend the timbers over the outside of all.

One of these methods is usually best for a particular hull, depending on the size of the frames specified, the amount of frame curvature and the skill of the builder. Either (a) or (b) is adaptable in our case, and we will treat (b) as it is slightly more involved. Method (c) is normally used for small open boats with very stout frames, and involves a great deal of time and trouble, while (d) has many obvious advantages, and is becoming increasingly popular for all sizes of craft. Normally, in this latter method, the ribbands are removed (perhaps by cutting in half) after the moulds have been dismantled, but they can be left in place and through-fastened to the planking as extra longitudinal stiffeners.

We must allow notches in the moulds to take the shelves; the notches will be the thickness of the planking plus the thickness of the frames plus the thickness of the shelf, in from the lofted lines; these thicknesses must be measured *on the plane of each mould*. If the shelves are to be made of two or three lengths scarphed together, each piece can be fitted separately, if necessary by steaming and bending round a former, or pattern, before trying it on the boat, and, having shaped it, re-steaming before permanent fixing. Of course, true permanent fixing cannot take place until the timbers and sheer planks are in, as they hold the shelf fastenings. If the shelf is to be in a single length, its shape will have to be determined by means of a spiling batten, in the same manner as the sheer strake of planking is marked out. This process will be described later in the planking paragraphs.

BUILDING A 3-TONNER

The term "shelf" is loosely used. A true shelf is wider than its depth, whilst a piece of timber doing the same job, and deeper than its width, should strictly be called a clamp. However, when only one member occupies this position, it is invariably called a shelf, whatever shape it may have, so we will call the one behind the sheer a shelf and the one along the cabin-top a clamp, to distinguish them. We will not fit the clamp until after planking: there is very little tumblehome in this region and the tops of the frames can be left protruding. By fitting the shelves before the planking, the whole structure is made very rigid and there is no likelihood of distortion during planking.

The ribbands are of about $1\frac{1}{2}$-in by 2-in. pine, screwed to each mould in the positions shown in Fig. 43. They should be single lengths, though there is no harm in using

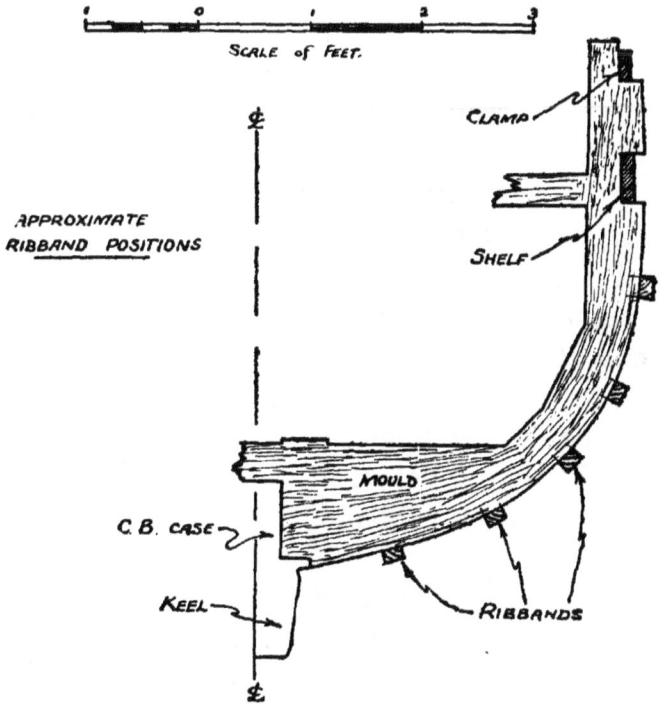

Fig. 43.—Approximate ribband positions.

many overlapping sections, but, to maintain the curvature, the overlap should extend over three moulds. The run of the ribbands is shown clearly in the accompanying photographs. They should not be strained into place, but steamed if necessary.

BENDING THE FRAMES

When well steamed, the 1-in. by $1\frac{1}{4}$-in. American elm frames should bend easily to the gentle and non-reversing curves of this hull, and there is no need to hurry over bending, as with the thin ribs of a dinghy. Having cut the frames to length, with 6 in. or 1 ft. to spare at each end, chamfer the inboard corners and number each rib for a corresponding number on the boat, pencilled in the correct position on the shelf, each ribband, and the keel, on each side of the hull. Examine each frame for flaws and bad grain; a few spare lengths should be prepared in case of breakages. When shaping the timbers, make sure that the grain will run parallel to the ribbands when viewed in cross-section, as bending is easiest with it this way. A certain amount of experiment may be needed to get the correct time for leaving them in the steaming-box, because they become brittle if left in too long, as well as if not long enough. It is a good plan to soak all the frames in cold water for several days. The frames which end at the keel, and notch into the hog-piece, are always easy to bend, but the long ones, crossing over the keel, may require bending round a former and tying across the tops with string before fitting into the hull. Fasten the frames at each ribband, if required. Either nail through hot timber into ribband or skew nail corner of ribband to timber (which does not leave a nail hole showing if varnish is to be used). In the former method the nails can be driven through previously drilled holes in $\frac{1}{4}$-in. or $\frac{3}{8}$-in. small pieces of soft wood before entering

BENDING THE FRAMES

the timber and hammered home hard wherever the timber tends to " stand off." These pieces can be split away and the nails withdrawn when the ribband is removed. No permanent fastenings should be put into the frames until they are dry and hard, although it is sometimes convenient to nail or screw the long ones to

Plate 12.—Ribbands and frames in place.

the hog-piece. This boat has no reverse curves to the bilge, rendering frame-bending a simple operation, but where such reverses occur, in other designs, it may be necessary to make formers for the worst of the frames concerned, and having let them set on the formers, take them off, re-steam them, and clamp into the boat. An alternative to double steaming in difficult cases is to use double (laminated) timbers to make bending easy, or else to saw the timber in half at the difficult place, and rivet it back together again after setting. The formers may take the form of simple cleats screwed to the floor, or a proper battened jig such as will be described later. Such a jig need not be accurately made, as packing can be inserted to correct. In any case, make the bend more severe than it need be, for the frame will spring out a little after removal, and it can be straightened a little by straining in a vice, though it cannot be bent more in the same manner.

PLANKING

With accurate lofting, no fairing up will be needed, and the frames will pull up to the planks where necessary as the fastenings are driven. Planking is the critical point in boat building, and round bilge planking is much more difficult than chine, so a full account will now proceed. First paint all the rabbets and outsides of timbers, and again lute all close surfaces as each plank is fitted.

We will start with the garboard strake. This will be the widest plank, and as it is nearly always under water and has little curve in the width, it can be made as wide as the timber will allow. A maximum width of 9 in. amidships is permissible. The use of a spiling batten has already been described, and the lower edge of the garboard strake must be picked up by tacking (with wire nails left protruding) a spiling batten into the rabbet after it has been roughly cut to the shape which, by trying it up once or twice, is found necessary in order that it may fit within an inch or so of the rabbet-line while lying without any distortion on each mould in turn. The pair of planks may be shaped together after spiling one side, but it is better to use the same batten and take spilings on

PLANKING

its back on the opposite side of the ship, thus assuring that both garboards will fit really well; if steaming is necessary, leave this until after shaping. The caulking

Plate 13.—Garboards and sheer strakes fitted.

opening on the seams will be $\frac{1}{16}$ in., tapering to nothing half-an-inch below the surface. Using $\frac{3}{4}$-in. planking, this makes the inboard $\frac{1}{4}$ in. a tight, perfectly fitting seam.

BUILDING A 3-TONNER

Through fasten (roved copper nail) at each timber at the top edge and the bottom edge of the garboards; the latter through the hog (or keel batten) also. One or two screws should go through the lower edges of the garboards in the keel rabbet between timbers; not in the same line of grain as each other. The necessary fillers must be fitted before the upper nail is driven.

Next fit the sheer strakes. In getting the shape of the sheer edge with a spiling batten, place the batten as accurately as possible in the position of the final strake: in other words, as near to the sheer as possible. Having transferred this line to the plank, a taper for the ends of the plank must be decided. If it proves about $4\frac{1}{2}$ in. wide amidships, allow it 2 in. at the stem and stern, and strike in a line for the lower edge to give a smooth curve. The usual way of ensuring a smooth curve is to scribe a circle whose radius equals the maximum width of the plank (see Fig. 44), then drop the perpendicular line A-A

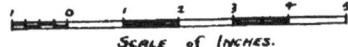

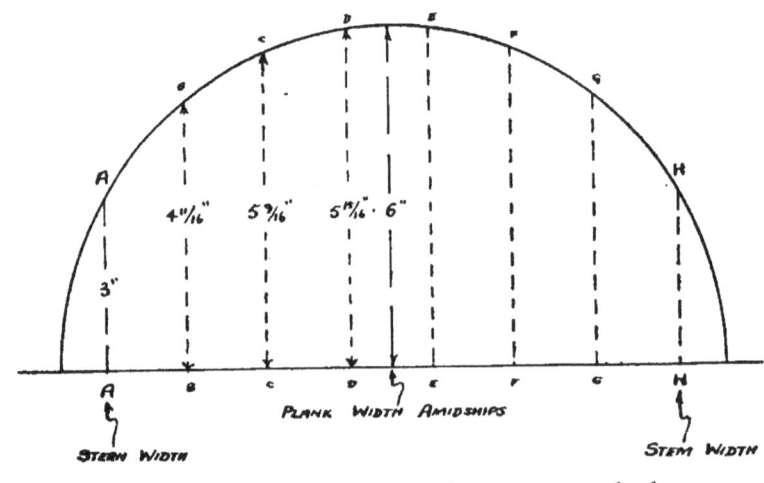

Fig. 44.—Plotting a smooth curve on plank.

PLANKING

of length equal to the stern width of the plank, and H–H equal to the for'ard width. Divide the space between them into the correct number of frames on the boat, and transfer all the distances to the plank. Run a pencil line through all the points, guided by a bent batten, and cut to the line. Plane up the edges to the correct bevel, with caulking chamfer, and the strake is ready to fit. The sheer strake will take most of the shelf fastenings, which should be driven next after fitting the plank. The breasthook and stern quarter knees which hold the ends of the shelves can then be bolted in.

Now that the two extreme planks are fitted, the widths of the remainder may be accurately determined, and a uniform space left to be filled in by the shutter plank, or final strake. As a guide, the approximate amidships plank widths, working upwards, should be 9 in., 6 in., 4 in., 3 in., and the remainder 3 in. until the sheer strake is reached, this being somewhat wider. The planks at the turn of the bilge should be narrower, say, $2\frac{1}{2}$ in., $2\frac{1}{4}$ in., 2 in., 2 in., $2\frac{1}{4}$ in., $2\frac{1}{2}$ in., and these should be "soled," i.e. made slightly hollow to fit the curvature and slightly rounded to the curve on the outside and will therefore need $\frac{1}{8}$-in. thicker plank timber. Don't forget to order accordingly.

The "laying off" of the planking in a pleasing manner is usually considered to be difficult for amateurs, but it can be tackled with confidence of success if the following mathematical method is adopted.

Fit one or two more planks next to the garboard so that nearly all the moulds (preferably all, where practicable) are reached by the ends of these strakes; the outer edges of these strakes to be parallel amidships, or slightly opening going forward. The lower edge must, of course, be spiled for, as previously described. Mark all the proposed positions for the seams on the mould which

BUILDING A 3-TONNER

has the greatest distance from the edge of the last plank fitted around to the sheer; usually the midships frame.

On a clear space of floor or preferably a sheet of plywood approximately 5 ft. by 4 ft., lay out this measurement as a straight line, full size, and transfer all the plank widths to it as shown at A–B in Fig. 45. Next, strike parallel lines C–D and E–F about 2 ft. on either side, which must be the exact lengths of the unplanked portions

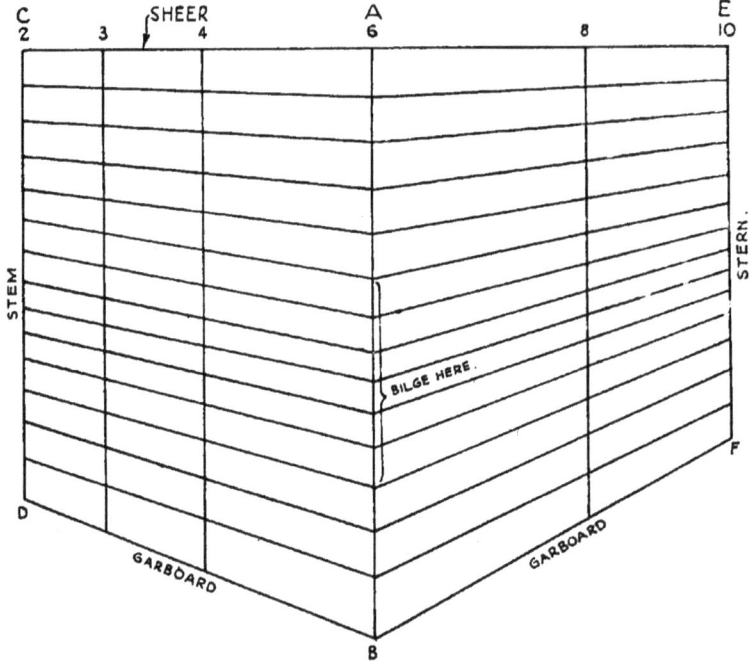

FIG. 45.—Planking layout.

of the moulds farthest for'ard and aft, respectively, reached by the planking already laid.

The same number of planks are now laid out on C–D and E–F, those on C–D (for'ard) being spaced with less difference between the plank widths at the extremities compared with the widths at the bilge. As the curve of the bilge continues towards the transom the narrow planks around the curve must gradually taper, so that

PLANKING

E–F must be divided up for these in proportion to the widths on A–B, the remaining planks gradually increasing in width upwards and downwards, as shown.

Plate 14.—Planking a double-ender—one ribband still in place.

The intermediate mould measurements are now taken, (with a tape) and *fitted in where the lengths come right* on the board as indicated for moulds Nos. 3, 4 and 8 ; the

127

widths of the various planks can then be picked up with dividers and placed on the various moulds to serve as a check as planking proceeds. If all the moulds were not reached by the lower planks, the degree of equalization of plank widths will be less, and in this case further planks should now be laid, working upwards until the extreme moulds are reached, when, if desired, the remaining planking can be laid off afresh on these moulds.

It is common practice to get the position of one or two bilge strakes first, by clamping them in place, so as to be sure that they are, as they should be, perfectly straight, though tapering towards the transom. This method is slightly more economical of timber and, if adopted, as will readily be seen, the principle previously described must be applied in two parts (1) below the bilge, (2) above the bilge.

ALTERNATIVE PLANKING METHODS

If the method of calculating plank positions, just described, is not adopted, an alternative is to mark up the proposed plank widths on the midships mould only and then run battens along the hull at each seam position and fair them in by sighting along them from all angles. All the seam positions should be marked before removing the battens.

Before proceeding with the actual laying of planks, it is now proposed to describe a method of getting the shapes of the planks which is very precise, yet requires very little extra labour and obviates spiling for each plank.

Firstly, very thin $\frac{1}{2}$-in. by $\frac{1}{2}$-in. battens of clean flexible timber are tacked lightly around the hull so that one is provided to represent each edge of a plank. In carvel build, the batten for the lower edge of one plank will touch the batten for the upper edge of the next strake

ALTERNATIVE PLANKING METHODS

below it, so that the surfaces which touch for the entire length will represent the seam between those two planks. For clinker build, all the battens will be separated, owing to the land widths, and they will still be arranged so that the upper edge of one batten represents the upper edge of a plank, while the lower edge of another batten (in

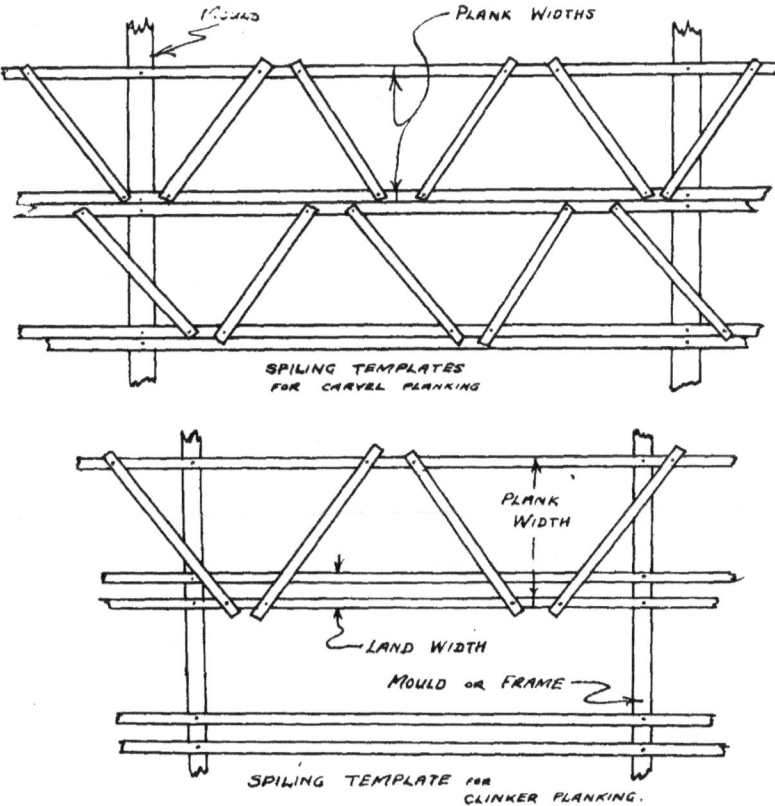

FIG. 46.—Spiling templates.

this case the second one down) represents the lower edge of that plank, as shown in Fig. 46.

Secondly, the two battens for each plank are tied together by means of numerous strips of $\frac{1}{2}$-in. thick wood, tacked across as shown in Fig. 46, to form a lattice girder. As each pair of battens is so treated, the resultant template is removed from the hull and laid on the floor. Its outermost edges represent the shapes for the edges of the actual plank, which can be marked immediately.

It will be apparent that all the lattice strips could be tacked into place on a carvel hull before removal of any of the templates, but for clinker construction, the templates must be removed separately to uncover the batten for the upper edge of the adjacent plank below. Using this system, the entire set of planks can be cut before any planking is commenced, and all the strakes will be of fine uniform appearance. The garboards may have to be spiled in the normal way, but elsewhere no spiling batten and monotonous divider work is necessary, and the same method can be used for other members in the vessel.

The battens used should cover the length of the hull in one piece, so the method is not quite so simple for big boats, and for clinker-built boats smaller than about 20 ft., LOA the battens representing the lower edges of planks must be raised from the moulds by a distance equal to the planking thickness, or the resulting spiled edge will not be exact. For larger vessels, the slight error occurring by fitting all these battens flush to the moulds is taken up by the spring of the timber.

PLANKING AND KEEL

To continue planking, work up from the garboards to just above the stringer positions, and fasten the planks which come opposite to the stringers very scantily, as the stringer fastenings will replace many of the normal ones. Then fit the binding strake, which comes below the sheer strake, and continue downwards. The ribbands must be removed as planks are laid in their place, and the floors can be fitted after a few strakes above the garboards have been fastened. Make light wood patterns

PLANKING AND KEEL

for the floors, then pick sound natural oak crooks for them. The keel bolt holes can be transferred from the hog-piece and bored separately, then the floors can be bolted in and wood screws driven through the planking into them. If suitable crooks are hard to locate, the floors may be straight-grained oak, if the depth at their centres is made appreciably greater.

Iron keel-bolts are essential to prevent local electrolytic action with the iron keel; these should preferably be galvanized, but lute or grease each bolt well before driving. These bolts can be ordered to length from many engineering firms, bought through a ship chandler, or made by a blacksmith. The heads of the bolts should be square, fitting into square recesses in the ballast keel. If a blacksmith makes them, get him to dip them in coal tar while red hot. The resultant surface will last longer than galvanizing. Ballast keel nuts are always set up on large heavy gauge washers; often square pieces of steel plate. Details of yellow-metal and other bolts are given in Chapter IX.

The upper rubbing strake which runs close to the sheer will be fitted after the deck, because it covers the edges of the canvas. It may be riveted or screwed through to the frames, and the lower one should also be fastened at this stage. English oak should be used for both these strakes. Timber for planking is not easy to choose, but makes a lot of difference to building. Avoid oak or elm, except, perhaps, for the garboards; Columbian pine rots very quickly in some waters (e.g. the River Severn). Local boat-builders will know the best woods for the waters near at hand, and should be consulted. Honduras mahogany is good for all planking, and teak is ideal, though, on account of its high cost, it is usually used only about the waterline, where many other timbers fail. Oregon pine, Douglas fir and Scotch larch are all good. Pitch pine is difficult to work, due to its resinous nature; it proves far better below the waterline than above, because, when dried by the sun, it loses its resin and becomes brittle.

The planking of the cabin-top sides may be left until this part is all framed up, with its deck beams in place.

CLINKER PLANKING

The order of fitting strakes for clinker planking is from garboard to sheer, and the chamfers or rebates at the lands (overlaps) should be made before steaming. Typical midships seams are shown in section in Fig. 47 (*a*) where a chamfer is required only on the one plank. This lap is common to all types of clinker construction, which differ only near the ends of the planks, where they are either gradually mitred or rebated into each other so as to produce a flush surface at the stem or transom. The mitred (or diagonal) method, near the ends of the strakes, appears as in Fig. 47 (*b*) and finishes as shown in a section of the stem in Fig. 47 (*c*) (left). This is the simplest and strongest, but is more liable to leak when the timber contracts.

The professional usually prefers the rebated method, the commonest type of which is shown in Fig. 47 (*d*), for a section close to either end of a plank, and which is terminated as shown in Fig. 47 (*c*) (right). This, however, is difficult for the amateur, who is advised to adopt the mitred construction, but to fit thin seam battens behind the laps, notched into the timbers, to make a more secure and leak-proof job. These battens need only extend for a short distance at each end of the hull where the outer surface of the planking becomes flush.

Clinker planking, except the garboard to keel joint, is not spiled for, it is *scribed*. The garboard outer edge is beveled to take the next strake before it is finally fastened

CLINKER PLANKING

in place, then the timber for the next strake is shored and clamped in position and a pencil run along the outer

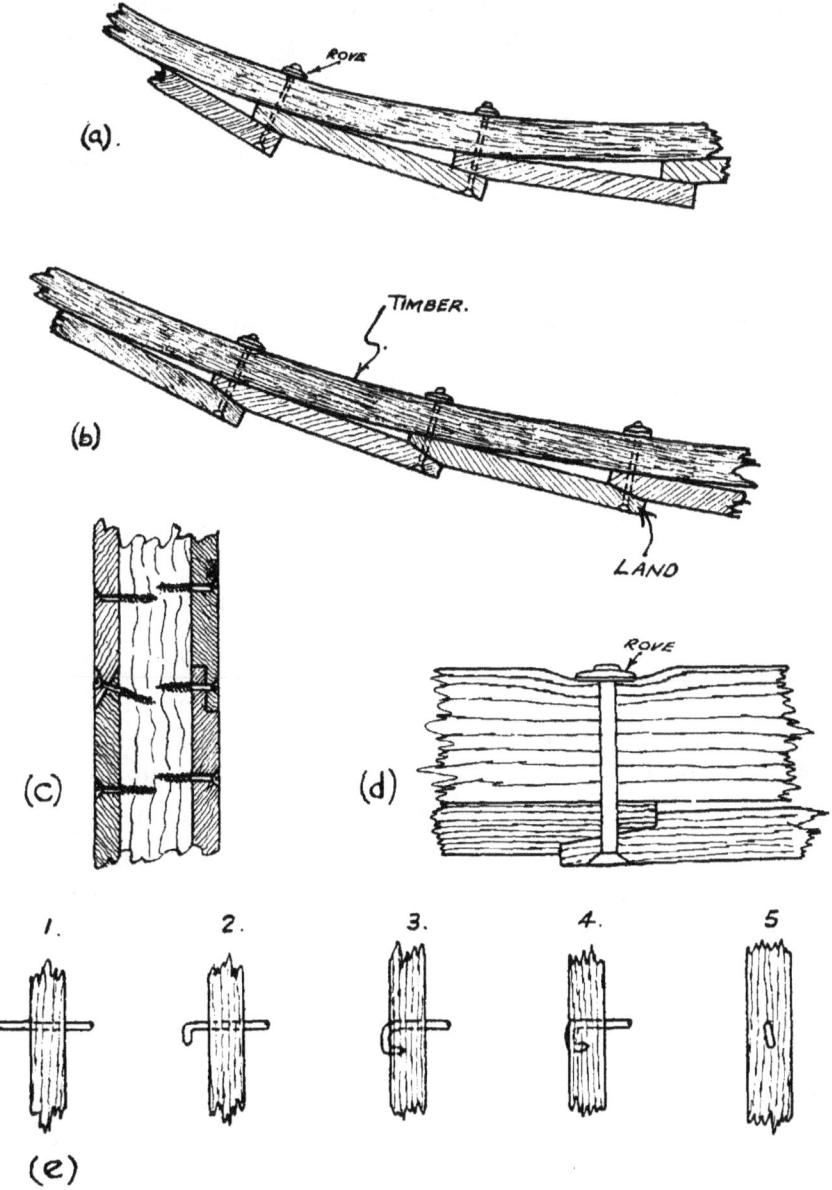

Fig. 47.—Clinker planking details.

edge of the garboard from inside: this is called scribing. The run of the outer edge of this strake is picked up from the marks on the moulds, which have (as a rule) been already marked off with the pre-decided plank widths. The extra width to allow for the " land " (fully one and a half times the plank thickness) must be included when sawing to shape. Clinker planking is very strong for its weight, with all the plank edges riveted up. Liquid marine glue is better than paint for luting between the seams, as it has more elasticity.

When riveting on roves the concave side of the rove must be placed next to the timber. Use a light ball- or chisel-paned hammer (about 6 oz.), and make sure that the rove is not quite flattened down when riveting is finished. (See Fig. 47 (*d*)).

The alternative method of stapling or clenching the nails which attach planking to the timbers is shown in Fig. 47 (*e*); the point should re-enter the timber at a slight angle to the grain. After the final hammering against the " hold-on " (or the face of a heavy hammer), the bent portion of the nail should be almost flush with the surface. If the nails are too long, make a gauge for the length required to protrude on the inside. Then cut off the point to the correct length with two diagonal cuts of the end-cutters (or side-cutters) to form a chisel point, and bend over, following the sequence shown. This method may be used for timber fastenings on all types of planking.

Deck Framing

Before removing the moulds, the clamps and most or all of the beams should be fitted. The latter are half-notched in depth into the shelf, and the main beams are also half-dovetailed in width.

The cabin-top framing may be completed and the side planks added ; all this, and the cockpit carlines and side deck beams, being quite straightforward carpentry. Iron

REMAINING JOBS

hanging knees and oak hanging and lodging knees are shown on the drawings, and, after these, the mast tabernacle supports, with their deck partners, and the for'ard samson post and partners can be made. The moulds will have to be partly dismantled before they can be removed, and as soon as this is done, the stringers may be fitted and fastened right through to the planking with copper nails, riveted on roves on the inside. The stringer breasthook (holding the fourth stem scarph bolt) and the cleats which attach the after ends of the stringers to the transom frame can then go into position.

Much interior fitting can be done before laying the deck, as space and light will be better. The decking is all parallel tongued-and-grooved boarding, with a covering of painted canvas. This produces a really tight and light-weight deck, which will last indefinitely if always kept well painted. There are no covering boards; the canvas is rolled on a thick luting of white lead paint and the edges are tacked to the topsides and later covered by the sheer rubbing strake. Any seams in the canvas on exposed parts of the deck must be neatly stitched by a sailmaker (unless the builder is fairly proficient with needle and palm), but to simplify laying the canvas, some joints will have to be plain lapped and tacked, or covered by a thin wood batten. The edge of the decking must be rounded off, to avoid a sharp corner on which the canvas may split later on, and quarter-round wood fillets screwed into the corners along hatch coamings should be set in white lead paint to seal those edges of the canvas. This, and the internal jobs, may be completed after launching, if preferred.

Remaining Jobs

The $1\frac{3}{8}$-in. thick rudder, shown in detail in the designer's drawings, is formed of two pitch pine or elm halves tied together by through-bolts. Oak bolster pieces thicken the stock and the gudgeon straps add to the strength. The pintle is a single long bolt, so the rudder could be shipped while the vessel is floating rather than before launching.

Again, the construction of the centreboard, of steel plate, resin-bonded plywood, or timber, is very straightforward from the drawings, and this board should be fitted before the upper part of the casing is made. The launching cradle should have been arranged so that no bearers obstruct the centreboard opening, and it can then be introduced from inside. Alternatively, the casing could be completed, and the centreboard fitted after launching by pushing it up from below.

We will not treat the subject of rigging, but a note on spar-making will not be out of place. If spars of the approximate size can be bought, a lot of time will be saved, as otherwise square stuff must be shaped, as required, by hand. A large vessel may use Norwegian spruce for her masts, in which case little shaping will be necessary. These poles merely need planing smooth after removal of the bark. When pine is used, the heart need not run through the stick; in fact, a piece of quartering is better, up to a few inches in diameter, and will not crack so readily.

Starting with a piece of square timber, the required tapers should be put on by the sawmill, and the next job is to chamfer the corners so that an octagon is formed. If the spar is a light one, this can be done by hand-guiding on a circular saw, followed by more corner chamfers made by drawknife and plane until the shape becomes round and can be smoothed off with plane and glass-paper. Variations occur, such as when a mast with hounds is being made, and a square has to be left at this point to enable the bolsters to be attached. Hollow spars,

LAUNCHING

except the square sort used in small racing craft, are rather too specialized for the amateur to tackle. Give solid spars a few coats of raw linseed oil before attaching metal fittings, and clamp the latter round the stick rather than bore holes through it to hold them. Such holes are not detrimental at the ends of a spar, provided they are small in diameter, but must never be put towards the centre, or in a mast near the deck.

Motor Installation

Unless a lot of estuary cruising is contemplated, or just very short spells of sailing, this boat is better without an inboard engine. A good outboard motor would prove a great asset, taking up less space, costing less, being more economical to run, and producing less smell in the cabin. Modern outboards are very efficient and reliable, the only drawback being the trouble of shipping and unshipping them. Should an inboard motor be necessitated, the quarter propeller installation recommended by the designer is quite simple to fit, but some amateur builders may prefer to leave the lining-up operations to a marine engineer. All the structural details of the shaft logs, glands and engine bearers are shown on the plans.

The details of the recess which is sunk into the deck adjacent to the transom and offset to port of the rudder stock to enable the outboard engine clamps to be made fast to the transom, are also clearly shown on the drawings, but let it be said that the transom should be heavily reinforced in this vicinity to check the transmission of vibrations from the engine.

Launching

On completion of all caulking and painting below the waterline, the hull may be launched, and all subsidiary work, such as rigging and interior carpentry, carried out

BUILDING A 3-TONNER

in a more convenient manner with the boat moored beside a quay or bank.

The length to which the launching ways must be extended can be calculated by means of a sketch and a few measurements, and if the time of a high tide or flood river level be chosen for launching, the least possible extension of the ways will be necessary. The only snag which sometimes occurs is the sticking of the sliding ways on the standing ways. Grease or tallow spread over the standing ways will decrease the friction, and crowbars may be used to start the cradle moving. To obviate any further jamming, a warp can be taken to a winch on the opposite bank ; or through a leading block attached to an anchor some distance off-shore ; or to a motor boat, and made fast to the cradle, so that the cradle can be dragged down the ways if necessary. Additional ropes should be rove for controlling the hull after it floats. By roping the cradle to the hull at the after end, it will lift from the ways as the vessel floats, and thus will not jam on the very end of the ways in the event of their being silted over.

CHAPTER IX
CONSTRUCTION OF A LARGER HULL—(1)

THE more common types of boat which can be built by the amateur have been treated in the previous chapters, but space has not allowed a description of all the other possible methods of construction. Therefore, before starting the final and most ambitious stage in boat building, the sawn-frame round-bilge type, we will survey briefly some of the methods we have so far ignored or just mentioned in passing.

COMPOSITE CONSTRUCTION

As far as many professional boat-yards are concerned, time and money may be saved by fitting a yacht which would normally have sawn oak timbers with bent steel angle frames, keeping all the remainder of the hull on conventional timber lines. Although a composite built craft usually has only her frames of steel, various other modifications may be adopted, such as steel deck beams, keelson, shelves and stringers. The floors are always of steel in these boats. The amateur would not find composite construction easier or less costly than wood, despite the difficulty in obtaining seasoned oak crooks for the frames and floors, but an established yard equipped with the machinery for heating, bending and galvanizing steel frames would benefit greatly by such methods. True, cold bending can be attempted, but the application of bevels is still awkward, and the only solution for the amateur is to get the complete frame assemblies made by a suitable factory, each composed of futtocks, floors, deck beams and hanging knees, so that they will drop in place on to the wood keel and work can proceed much as for a timber vessel. Great accuracy in lofting is required because fairing-up is impossible. Fastening is quite simple, the planks being attached by one through-bolt or rivet at each alternate frame and wood screws driven from inside in all the remaining places. If the frame assemblies are painted with red oxide or bituminous paint after galvanizing they should last as long as the parts which are of timber.

All-Metal Construction

Steel or one or other of various light alloys, such as " Birmabright," is not infrequently used for almost the entire construction of small craft. For dinghies and similar hulls, the skin sheets can be shaped over a wooden mould, then welded together, and the floors, beams, stringers and gunwales fitted after lifting the skin off the mould. Bigger vessels are plated up over a conventional metal framework of ribs and keel. Skin seams may be butt-welded, lap-welded, riveted, or riveted and welded, a flush surface being best. A metal yacht has many advantages over her time-proven sister, but internal condensation and the difficulty of fixing ceilings and cabin fittings are defects hard to overcome. Another drawback is the comparative ease with which they unavoidably get dented, often in places where even the removal of the worst of the defect is difficult.

Sawn Frame Construction

However attractive a metal hull may appear there is a strange fascination about working in wood, which causes enthusiasm to increase as a hull takes shape, so we will ignore the details of metal hulls and carry on with the type of construction which is used for the larger sailing vessels and for the stoutest deep sea yachts above about 35 ft. LOA. Many boats of this type have a few bent

SAWN FRAME CONSTRUCTION

frames between the sawn ones, and although the idea is satisfactory for day cruisers and racing yachts, and makes for lightness and economy, its adoption is a doubtful asset in craft with planking greater than 1 in. in thickness. No special details of this bent and sawn frame method are necessary; it follows common-sense lines, deduced from a brief study of the two separate methods involved. A really large ocean cruiser is technically no more difficult to build than a bent frame pram dinghy, but it will take much longer.

First, the importance of choosing a good design must be realized, as this will control the success of the result. The conventional modern type of fast ocean cruiser, suitable for racing, is a somewhat stronger version of a modern racer, with a little less overhang. Her performance in rough weather cannot always be boasted about by her owner, and at times she may be very uncomfortable and difficult to handle. Although an amateur can tackle such a hull, this type of yacht is not recommended for a first attempt. Building is intricate when long overhangs, rockered keels, external ballast and long counters are adopted. On the contrary, the older type of cruiser is comparatively easy to build and provides nothing but encouragement. Furthermore, the sort of yacht which will be chosen as a building example will prove a more satisfactory deep-sea cruiser. She has only one disadvantage over the modern type, and that is speed to windward, but surely this is more than counter-balanced by her better accommodation, easier motion, drier decks, lower initial cost, easier graving, and safety in handling for amateur crews?

Having decided the type of hull simplest to build and also best for the owner's requirements, a list of probable dimensions can be drawn up and the shape of the stern decided. The transom stern is, for an amateur, the

CONSTRUCTION OF A LARGER HULL

simplest to build, though not ideal from the point of view of the behaviour or appearance of the boat. The canoe stern, which is also simple to build, has a great number of admirers both from the æsthetic and the " sea kindly " point of view, and a yacht on the lines of a Colin Archer-designed Norwegian pilot cutter or lifeboat ketch would therefore suit the amateur. The sawn-off counter and the full counter are somewhat more difficult. I believe it was Arthur Briscoe who said, " All sterns are good sterns if properly designed." This is undoubtelly true, but the full short counter stern is claimed by many as best for cruising in British waters. In the opinion of the writer this type definitely has the best appearance of all, and therefore in order to treat all the constructional details of a full counter an example of this type of boat has been chosen for building description. Her lines drawings are reproduced in Fig. 48. She is modelled on the lines of a Bristol Channel pilot cutter, and although she may seem too large for amateur construction, all the details required for a smaller or larger vessel are embodied.

The amateur should, perhaps, be warned that the *dimensions* of a design must not be departed from except to a very minor extent. An attractive design cannot be scaled down in exact proportion simply because it is too large. The resultant hull might not prove a successful sea boat.

Labour Arrangements

The question of labour becomes far more important now that we are dealing with a bigger boat that will take too long to build unless a sufficient number of hands is employed.

Paid labour would solve all these problems, of course, but should only be used as a last resort. Labour is the

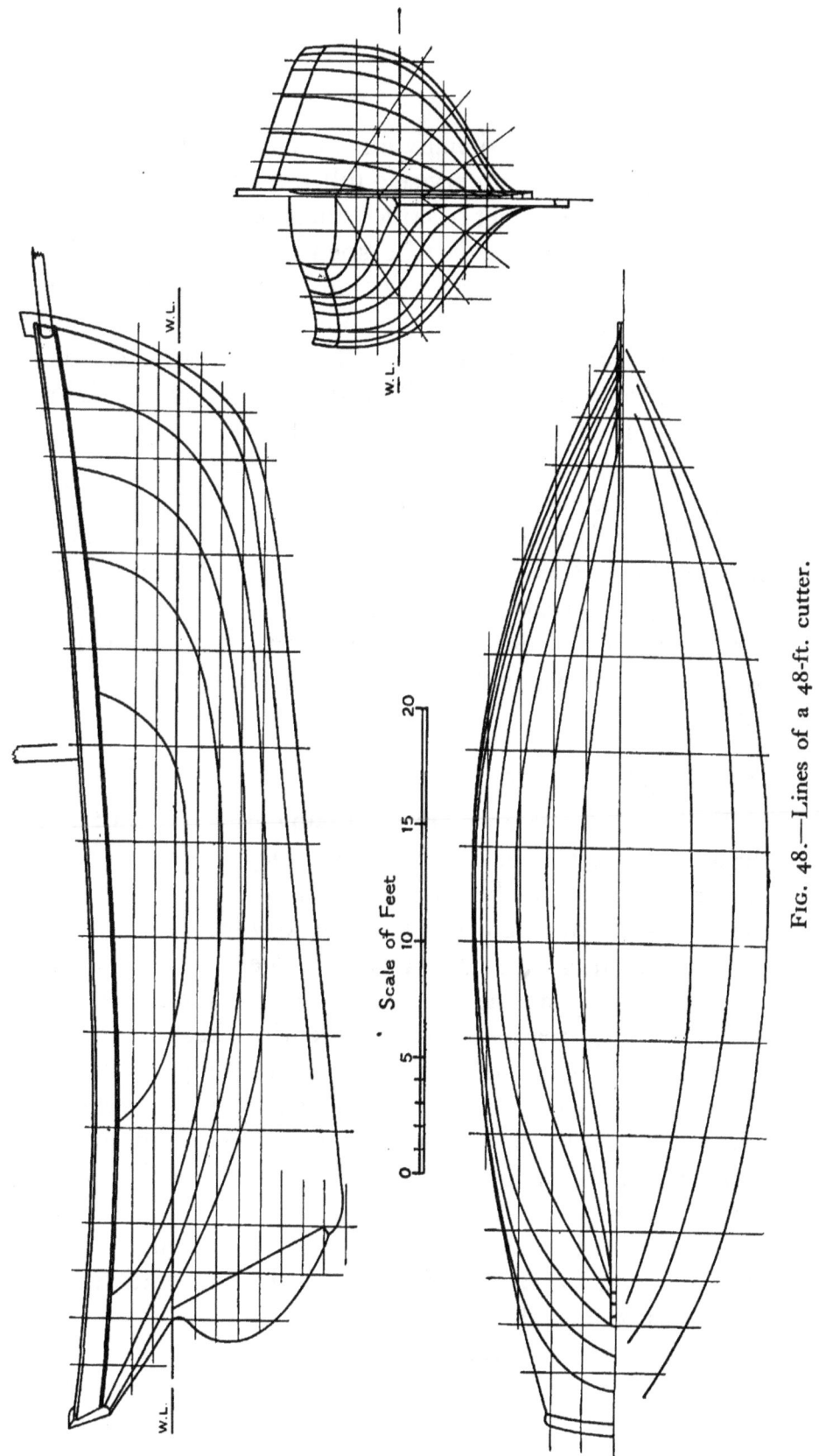

Fig. 48.—Lines of a 48-ft. cutter.

major cost of a professionally built yacht, and most would find it impossible to supervise a full-time man. An alternative would be to get a skilled man to work while one is also on the job, so that he spends his leisure in this manner for an agreed remuneration. But, if possible, try to keep the whole thing to a team of amateur enthusiasts, and probably the assistance of a few keen youths can be obtained for occasional small gratuities. Only a little of the work is really skilled, so that undue trouble need not be taken to get only the most ideal helpers.

When tackling a big boat, one should first get the constructional details passed by the B.O.T., and then arrange for a Surveyor from Lloyds to visit the building site every few weeks. The fees are very reasonable, and by this means one can obtain a Building Certificate which would prove invaluable if, later, one wished to sell the boat.

The Building Site

Yachts of the type we are discussing are very stout and heavy for their size, and although they could be built and transported to the coast, such a worrisome business would be eliminated by erecting on launching ways similar to those described in Chapter VII, but of proportionately heavier construction, of course. Again, there is no reason why the entire keel assembly and all the sawn frames should not be prepared in the backyard, or at some adjacent sawmill or carpenter's shop where all the tools and machines required are available, and then transported to the launching ways for erection.

If the prospective builder is fortunate enough to be friendly with the owner of a boat-yard who would render casual assistance, the best way of all is, of course, to build as near to it as possible.

THE TIMBER ORDER

THE MOULD LOFT FLOOR

A complete lofting of all three sets of lines is essential for a sawn-frame boat, as an inaccuracy will render fairing-up a very prodigious or impossible task. The mould loft floor can be made in two separate halves, with sufficient overlap to enable the lines to be faired, but everything will be much easier if the whole thing can be laid down on one floor. The best location is a smooth wooden floor with a roof over it, but as the amateur builder is slow, and will have to leave his lines down for a considerable period, it is rare that such a perfect floor can be found near to the building site. Therefore, an outdoor floor may have to be used, formed either of many strakes of flat boarding, cleated together at the ends and sides, or of a $1\frac{1}{2}$-in. layer of concrete on a hard prepared piece of ground. This surface will not crack if always walked on with rubber shoes and the lines can be drawn on with red or black surveyor's chalk, proving very durable if the floor is kept covered with a tarpaulin when not in use. After building, the slab can be broken up with very little trouble. Most builders loft to the inside, not to the outside of the planking, as the patterns, or moulds, can then be lifted straight off. This is the simplest way when the table of offsets is given to the inside of the planking, but when outside offsets are provided it is better to deduct the planking thickness on the loft floor by striking in new lines on the body plan the correct distance in from the existing ones. Because this distance varies, the above system is far easier in practice than recalculating the entire table of offsets.

THE TIMBER ORDER

Accurate quantities of all the timber required are difficult for an amateur to assess, as so much of the framing is curved and involves wastage in cutting. Therefore,

CONSTRUCTION OF A LARGER HULL

although a list of all the straight stuff may be compiled and converted to the exact sizes by the sawyer, patterns are needed for the curved pieces, and the natural crooks to make them will have to be found by a thorough search through many piles of slabs. Of course, a separate pattern for each small member is not necessary, as a general shape will probably give the approximate curvature for ten or fifteen frame futtocks. Likewise, many of the floors are very similar in shape and are therefore easily chosen. Special patterns must be made for the stem and apron crooks, and also for certain of the knees. Timber merchants catering for boat building used to keep big stocks of seasoned oak which were used in rotation according to the period of air seasoning. Such stocks are not so easily found to-day, but kiln-drying and chemical seasoning are speeding things up, and are quite satisfactory when expertly supervised. No planing will be needed on the straight stuff before it is shaped and fitted to the boat.

If oak is to be used for any of the planking it will probably have to be specially cut. Although the planking will not be needed until a fairly late stage in building, it should be ordered at the start to ensure its seasoning. Planking timber should be slash-sawn to facilitate bending, and this means cutting from the sides of the log to keep the grain within about 45 deg. of vertical, as shown in Fig. 15. Although a slash-sawn plank will bend more easily than a rift-sawn one, it is more prone to warping while drying-out after sawing, and in this manner it may take up a permanent curved section, or even split along the centre of the convex side. However, if the timber has been well seasoned previously, only a slight warping will ensue, and this can be very useful, as it reduces the amount of soling necessary to make the plank fit the curve of the frames. Some builders always obtain their solings in this manner, or by increasing the amount of

START OF BUILDING

warping by soaking the convex surface of the plank with water. This method is not recommended as it may mean that the plank is under tension on its fastenings when it dries out. The other planking, of soft wood or teak, can be ordered later, when needed. Unless the sawmill has a collection of sharp crooks suitable for knees, these will have to be collected slowly. The limb crooks which make knees are most frequently not brought in with the log, but are left in the forest as top timber for firewood.

A reduction in cost may be effected if all the timber can be ordered at one time, and there is then no likelihood of a hold-up due to a certain order not arriving at the required moment. Where the timber should be stored is a matter depending on the facilities for sawing it and the available space ; the sawing of various members will be reviewed later.

Start of Building

After lofting, all the patterns can be made. Next scarph all the members of the main framing together to

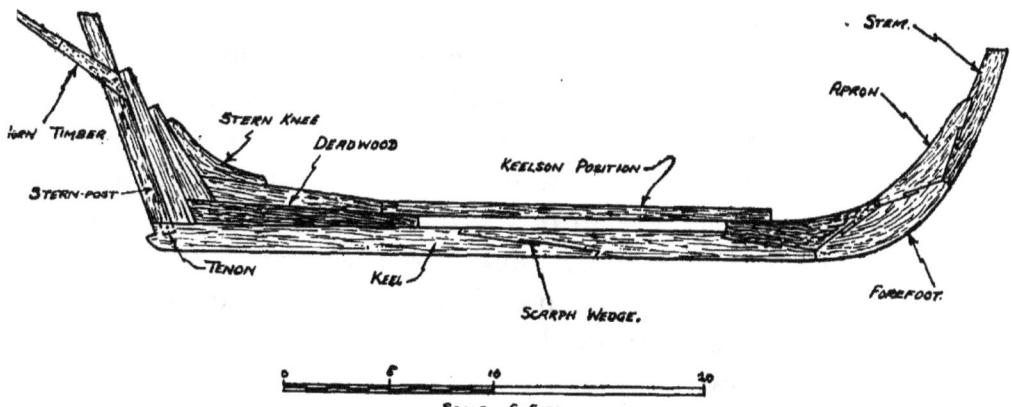

FIG. 49.—An example of scarphing the main members of a 48-ft. hull.

form the structure shown in Fig. 49. Of course, the arrangement of these members is arbitrary and may be altered to suit the shape of certain available pieces of

CONSTRUCTION OF A LARGER HULL

timber. All the scarphs will be shown on the mould loft floor, but patterns will only be required for the curved pieces, measurements with a steel tape and large pair of dividers sufficing elsewhere. Such dividers can be made from two pieces of hardwood, bolted together to hinge at the one end and tapered, drilled and fitted with a sharpened nail for each " point." With a length of about 3 ft. they will open to any of the distances required for transferring the scarphs and moulded dimensions from the lofted lines to the actual timber, and the steel tape can be used for all the measurements greater than about 4 ft. Incidentally, these dividers are ideal for lofting the lines. A steel tape can be laid along the baseline, and all the offsets taken from the tape with the dividers and transferred to their respective stations.

The scarphs should preferably be made self-locking, and wedge scarphs, such as that in the centre of the keel, are simple to cut besides being almost the strongest. There is little danger, however, in employing plain scarphs if well made, adequately long, glued with cold-water glue and well bolted. All those bolts which do not interfere with the floors can be driven on erection of this framing, but where these are insufficient, temporary bolts can be used. The keel is very stout, being 15 in. by 8 in. in section, and should any difficulty be experienced in obtaining the seasoned oak in this size, there is no objection to laminating this member into three separate horizontal layers, with the scarphs in each layer well shifted, and the whole structure thoroughly bolted. The central scarph in the keel shown in Fig. 49 is actually a weak point, but the scantlings are so generous that it would not be detrimental, and this scarph is covered by the keelson above it. The keelson could also be laminated in a similar manner, so that all the pieces are easily obtained and light to handle. With the advent of good, simple, resin glues,

START OF BUILDING

such products could well be used between the laminations, with considerable addition of strength.

If no glue is used, dowels or tenons should be set between all the surfaces, to eliminate any tendency for sliding or working. Setting these dowels usually means boring into each adjacent surface in a very precise manner, or chiselling a square mortice where a wooden tongue is used. But a far simpler method, that is almost equally efficacious, is to bore straight through the seams sideways, as sketched in Figs. 50 (*a*), and fit a not-too-tight hori-

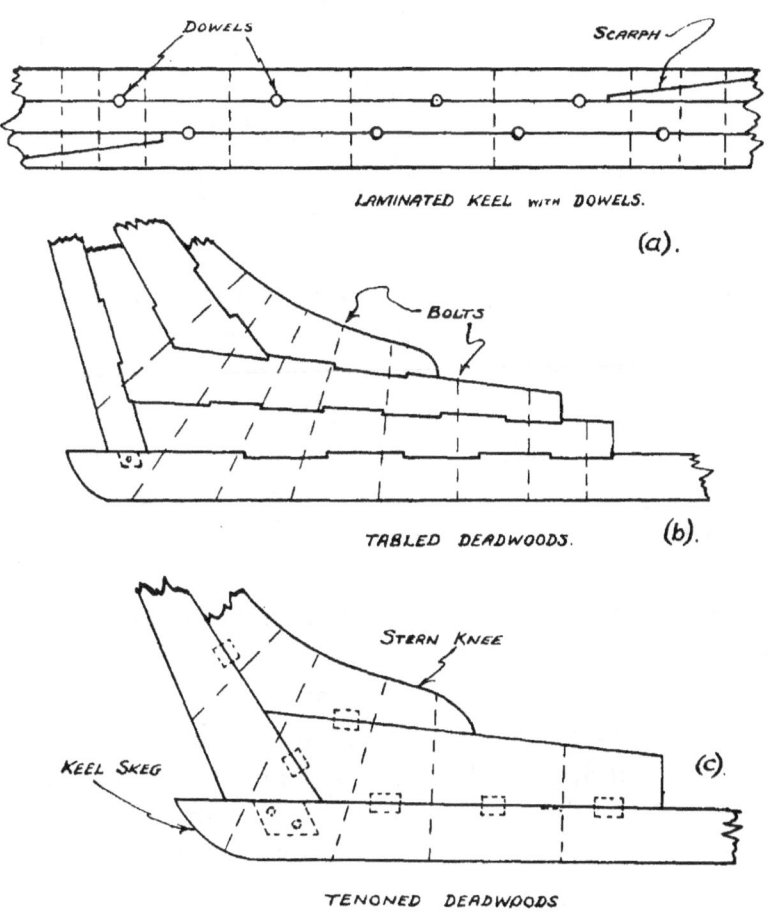

Fig. 50.—Locking adjacent surfaces.

CONSTRUCTION OF A LARGER HULL

zontal dowel into the hole, similar to a stopwater. When the two members concerned are tightly bolted together, these dowels will relieve the bolts of shear stresses, though they will not prevent any side slipping of the timbers : a tendency which is not likely to occur. The deadwood members can be tabled together, as shown in Fig. 50 (*b*), a very strong method, but also a difficult one. The more common tenon method is represented in Fig. 50 (*c*).

Although the keel, stem and sternpost are sided 8 in. in this boat, the deadwood should all be 5 in. to avoid cutting the absurdly deep planking rabbet which would otherwise be involved. The only objection to this is that a dirt-collecting crevice is formed between the deadwood and the planking which would need to be filled with cement, marine glue or other compound.

The material used for keel bolts is quite an important point and is worth a few lines of discussion. A highclass job or a boat which is to be used in tropical waters necessitates copper sheathing, so that some copper alloy must be used for bolts to avoid electrolytic action with salt water. The diameter of the bolts depends upon the alloy, a weak metal like 50 per cent. brass needing perhaps twice the diameter of Tungum alloy, the latter having a maximum tensile strength of about four times the former. Naturally, small bolt holes are less liable to weaken the timber, so that special high stress alloys should be used for all keel bolts. Stainless steel can be used, but its strength is very little greater than Tungum alloy, or Delta bronze, and it is much more difficult to thread and cut, though there is little difference in cost. The bolt metal can be obtained in long lengths of extruded bar, cut to length on the boat and threaded each end for nuts. A nut on the underside of the keel is easier to make than a head and less damaging to the metal, and it can be riveted over before driving. Some builders use clench bolts for scarphs and other keel fastenings, which are usually of brass or copper riveted at each end over a large rove or washer. This method is good, but not recommended for the amateur. Drift bolts are used a great deal by professional builders. Galvanized iron ones are best, and they are commonly cut from a length of bar by the blacksmith, who hammers a slight taper on the entering end. A washer is put over the bolt, and it is driven right to the end of its prepared hole. The burred outer end is cut off if necessary, and this end is then riveted up over the washer. Occasionally a screw thread is put on the outer end. A washer and special threaded cap are put on, and on completion of driving, the cap is removed, and replaced by a nut which is tightened home with a spanner. Drift bolts hold much more firmly than most amateurs imagine.

Iron fastenings have proved their long life and strength, and if well galvanized or hot-tarred they are really superior to yellow metal. But the quality of iron varies, and surveyors treat such fastenings with great suspicion; they lower the selling value of a yacht considerably, and tend to affect the behaviour of the compass. Copper sheathing should not be used on an iron-fastened boat, because of galvanic action, but zinc sheathing is quite satisfactory, providing the iron plank fastenings are well galvanized, and is far cheaper than copper. Galvanized iron sheathing would not be very successful, for the nail holes would puncture the zinc surface, and again, the behaviour of the compass might be seriously affected.

If a long auger cannot be borrowed to get right through the deadwood bolt holes, a piece of rod must be welded into the shaft of a normal-sized auger to lengthen it. The piece of steel used to extend the auger should be rather less in diameter than the shaft of the drill, and may be finished to fit a carpenter's brace in lieu of the usual hole

CONSTRUCTION OF A LARGER HULL

to take a wooden bar. The "shell" type of bit is cheapest and is quite satisfactory. Either this or the bull-nosed pattern (without feed screw) will guide accurately through a long hole; the spurred types are more likely to wander. If a brace is used choose one with a large throw. If an electric hand-drilling machine and power supply are available an extended morse drill may be used, but there is a tendency for the grooves of the drill to

Plate 15.—Sternpost and keel, showing scarphs.

clog and, if this happens, the drill should be withdrawn at intervals and the grooves cleared with a small screwdriver.

The plank rabbet is best cut before the framing is raised on the stocks, and a mallet and sharp 2-in. wide chisel are the best tools to use. The method of getting the depth between the rabbet and bearding lines has already been described, and some time may be saved by boring a series of holes to the required depth at the after

START OF BUILDING

end of the keel (where the rabbet is square and deep) with a brace and $1\frac{1}{2}$-in. centre-bit.

If work has so far been accomplished in the backyard, do not dismantle anything until all the frames have been fitted to their notches along the keel. The counter timbers may also be made before erection on the stocks, so that work should go with a swing as soon as planking is started.

CHAPTER X

CONSTRUCTION OF A LARGER HULL—(2)

SAWN FRAMES

OUR example has a full set of double-sawn frames, spaced at 16-in. centres, the making of which will shortly be described. This arrangement gives the strongest possible hull, and is the same as that used in the bigger wooden vessels. One small difference exists between our example, and, say, a Brixham trawler. The old West Country builders, who used their eyes and experience more than drawings, set all the timbers at right angles to the keel, instead of to the waterline.

Their way is in some respects easier to build, because the keel notches for the heels of the timbers are all square to the keel and the floors sit flat on top of the keel, with no bevel. But unless the keel is laid horizontal on the stocks, the frames will not be perpendicular, and if the ways are to have any declivity, the stern of the boat will then be inconveniently high above the ground. If the waterline of our example is arranged horizontally, the keel will have a declivity about the same as the ways, and the frames can all be set dead plumb, which is ideal for building.

Lines drawings with the keel shown square to the frames are difficult to comprehend when one is used to the more usual system, wherein the waterline is parallel to the baseline.

Sawn frames can be double futtock, partly double, or single, a view of each being shown in Figs. 51 (*a*), (*b*) and (*c*) respectively. Our example has all her frames double, but many stout vessels, such as the Morecambe

SAWN FRAMES

Bay prawners, oyster smacks and various yachts, have timbers doubled only up as far as the stringers. Others have a few frames amidships full double, with partly double ones for the remainder, or even with a few single frames in the bows, where there is little curvature.

For a deep-sea boat, such construction is often false economy, and a yacht may need more strength in her

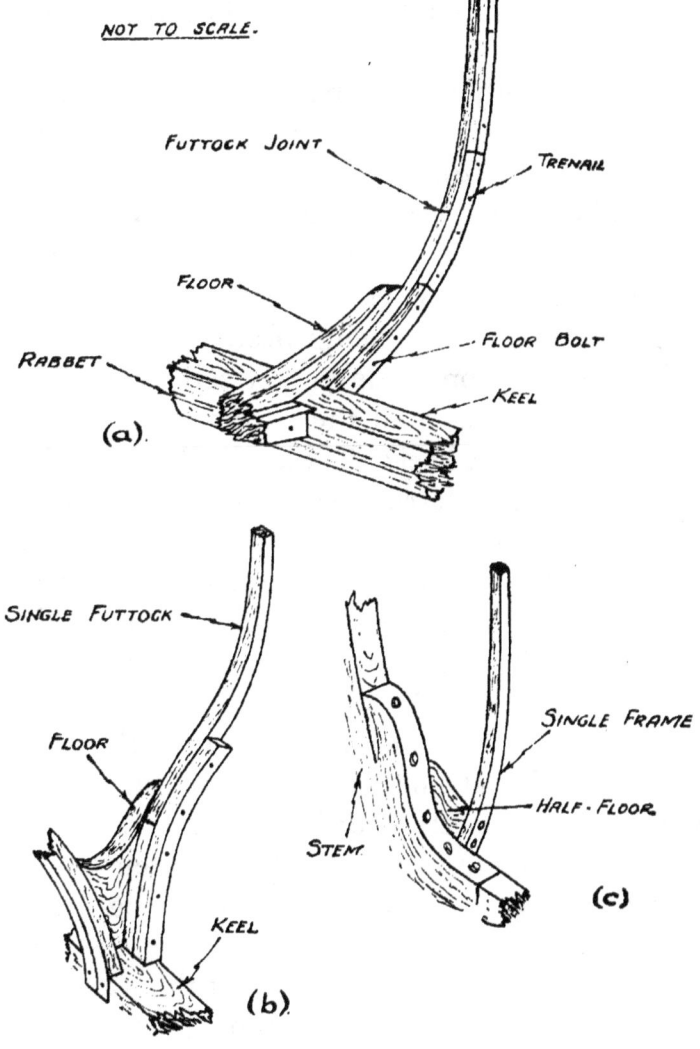

FIG. 51.—Types of sawn frames.

bows than amidships, so we will conclude that a full set of double timbers is best for the type of craft in question. Each futtock is sided 3 in., giving complete frames sided 6 in. and moulded from $4\frac{1}{2}$ in. at the heels to $3\frac{1}{2}$ in. at the heads.

Therefore, all the crooks used should be 3 in. thick, and great care and trouble should be taken in choosing the best crook for each futtock. Long futtocks are better than short ones, providing they follow the grain of the wood everywhere, and as each crook may make several futtocks it is usually best to make each side of each frame identical, preferably of two futtocks fayed to three, as shown in Fig. 51 (a).

Before any frames can be made, or the crooks chosen for them, a pattern must be lifted for each frame from the mould loft floor. These patterns can be of very light timber, about $\frac{3}{8}$-in. pine being best, with gussets across the joints so that one side is perfectly smooth. This smooth side must have all the surmarks upon it, which consist at least of the keel side line, the diagonals (and any waterlines used for referring bevels to), and the sheer-line. Also, the bevel at each diagonal or extra point can be marked at that point for speedy reference. The bevels can be measured to a degree from the loft floor lines, but can alternatively be calculated and actually drawn to angle on a bevel-board for each frame. The bevel is slight amidships, but increases towards the stem and stern. Especially at the bows, the timbers sometimes have so much bevel that cutting becomes awkward and wasteful. Therefore, cant frames are often used at the first four or five stations, as shown in Fig. 52. Cant frames are not only easier to saw; they are stronger, as they come in line with the direction of the water pressure on the hull, and the plank fastenings can be driven squarely through both the planking and the frames.

SAWN FRAMES

Although the fitting of the floors is not quite so straightforward, it can be made amply strong. If cant frames are decided on, the new stations must be struck in on the loft floor, and the moulds lifted from these.

In matching the futtocks to the moulds, it must be remembered that the mould represents the widest part, with the bevel turning inboard from it, and the trend of the line separating heart wood from sap wood must therefore be watched. The opposing futtocks from one frame can generally be cut from the same crook, and often

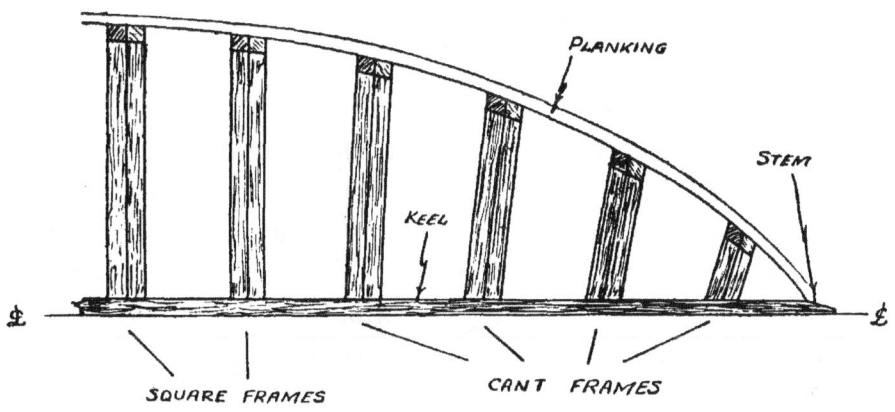

FIG. 52.—Cant frames in the curve of the bow.

another pair of similar ones, so only a single line need be cut between such pairs.

A band-saw is the ideal machine for this job, and is very quick, but if one is not available, the amateur will either have to get all his futtocks cut by a sawmill (after marking them all out on the timber and listing the bevels), or else resort to a pit-saw or other manual device; very slow and discouraging. Bevels are always applied during sawing, and, even with a canting-table band-saw, the merging of one bevel into the next is not simple, but, if well done, the amount of fairing-up will be negligible.

CONSTRUCTION OF A LARGER HULL

All the futtocks for each frame side are then fitted together by getting them to coincide with the mould and fastenings are driven to hold the whole assembly rigid. As all the plank fastenings in a well-built yacht should go right through, from planking to inside of timbers, these bolts eventually hold the futtocks firmly together; so, in assembling the frames, only sufficient fastenings to hold everything rigid enough for erection and planking are needed.

Trenails of $\frac{3}{4}$ in. diameter hardwood are ideal for this, arranged as shown in Fig. 51 (*a*), although clenched copper nails would do equally well. The trenails must be wedged at both ends, and are ideal because they will not obstruct the plank fastenings. These inter-futtock fastenings tend to lose some of their importance as soon as all the plank through-fastenings are driven, so it would matter little if one of the trenails was touched by the auger during planking. More will be said of trenails on page 169. All close surfaces should be well luted with paint before fastening, but tar or creosote will do the same job; in any case, all frame surfaces should be treated with hot creosote or Solignum before erection on the keel. Fair-sized limber holes should be cut at the lowest point in each frame, to allow all the bilge water to flow aft to the pump well. If the limber holes are too small, they will clog easily, but a length of small chain can always be rove through them all to facilitate cleaning.

FLOORS

The port and starboard halves of each frame are made separately, and then each pair is joined by a floor. The best floors are made from natural oak crooks, but perfect crooks are very difficult to find for these long awkward members, and that is the reason why steel, iron or bronze floors are so often used. Wood floors are best because

FLOORS

they have more resilience and they can be shaped and fitted in the yard. Patterns have to be made for metal floors, and sometimes they have to be brought from the foundry for marking and boring and then sent away for galvanizing. A mixture of wood and metal floors should be avoided at all costs.

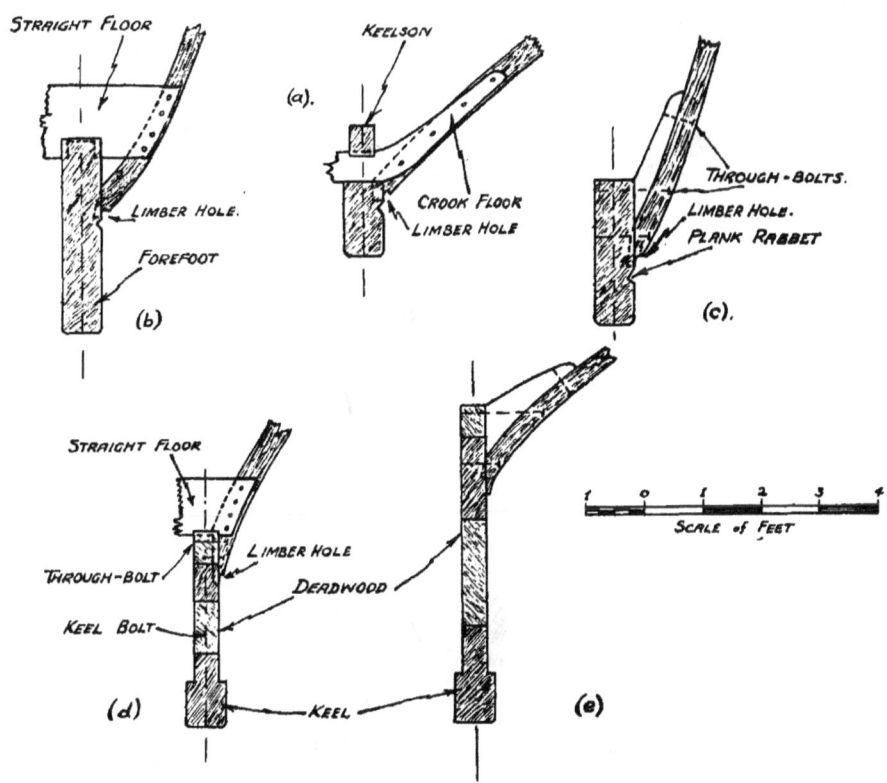

FIG. 53.—Five examples of oak floors.

Perfect crooks cannot be used for every floor; only the ones below the keelson can be so, a typical example being shown in Fig. 53 (a). The two for'ard ones in Figs. 53 (b) and (c), as well as the after ones in Figs. 53 (d) and (e), are built up and do not require natural crooks. Metal floors can be of the plate type, vide Fig. 54 (a), or

CONSTRUCTION OF A LARGER HULL

forged, as in Fig. 54 (*b*), and either can be made of wrought iron, mild steel or bronze.

The ferrous floors must be treated, by galvanizing, cadmium plating, or dipping in molten brass spelter. The latter process is little used, but it enables yellow-metal fastenings to be used and is far more efficacious against rust than any other process. Forged floors must be very accurately made, to fit with a minimum of frame adzing, and the fastenings are awkward because they are

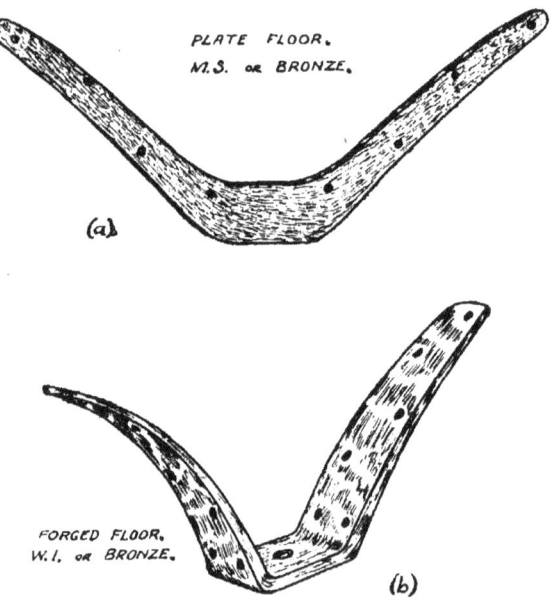

FIG. 54.—Two types of metal floors.

liable to coincide with the bottom plank fastenings. These floors are used mainly in conjunction with a stout ballast keel, as no keelson can be fitted above them.

Where metal floors are used with a keelson, the best arrangement is that shown in Fig. 55. Each floor is composed of two $\frac{3}{8}$-in. mild steel plates, bolted to either side of the frame, and therefore 6 in. apart, with a cap of the same metal across them to take the keelson. A hole about 2 in. in diameter in the cap allows the keel

KEEL NOTCHES AND KEELSON

bolt to be bored for and driven after the keelson is in place, without fear of striking metal with the auger. Timber floors have to be bevelled where they sit on the keel, and the full crook floors in our example are to be fastened to the frames with four $\frac{3}{4}$-in. bronze or $\frac{7}{16}$-in. stainless steel bolts.

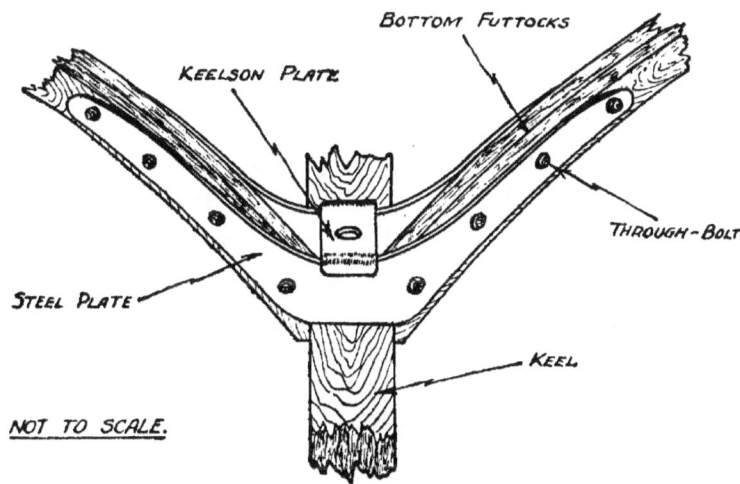

FIG. 55.—Plate floors with keelson plate.

KEEL NOTCHES AND KEELSON

If all the work so far has been carried out in the backyard, the notches for the heels of the timbers, in the keel, can now be cut, and each complete frame lifted into place for fitting before anything is transported to the erection site. The heads of the timbers must, of course, be linked together with temporary spalls, or cross ties, just below the shelf position.

Some yachts are built with no keel notches, relying on the floors to take all the strain should they ever strike a submerged bank. This is very bad practice, and unless the heel of each frame rests in a notch, the garboard rabbet fastenings will be started sooner or later, and leaks will occur. The shapes of the notches are shown in

CONSTRUCTION OF A LARGER HULL

Fig. 53, for typical stations. The method of erecting the main framing on the stocks is exactly as described in Chapter VII, and when this is done, the frames can be fitted permanently and shored accurately in position. When the side ribbands are fitted, many of the shores can be removed, so that only every fourth frame or so remains supported from the ground.

The keelson is sided 5 in. and moulded 6 in., in a single length of pitch pine or oak, joggled 1 in. over each full floor. As seen in Fig. 49, the keelson is bolted to the stem apron and deadwood at its ends, so that when the keel bolts are driven, each passing right through the keelson, a floor, and the keel, a very rigid girder is formed as the backbone of the boat. As no ballast keel is fitted, the strongest keel bolt arrangement is one $\frac{5}{8}$-in. stainless steel bolt at each floor, providing a small hole in the timber with the maximum required stress when the bolt is tightened.

BUILDING EQUIPMENT

Various tools and materials must be collected at the ways before erection can take place, such as scaffolding trestles and planks, assorted scantlings for shores and braces, a shed or two to lock away tools and small materials, bolts, nails, clamps, ropes and tackles.

A derrick will prove to be essential and should have a very long jib, with four guy tackles and lifting tackle to stand about half a ton. Two larch poles fished heel to heel and stepped into a shallow hole in the ground, provide a good jib, and a small winch is best for lifting.

In all cases try to build adjacent to a friendly boat-yard, from which many tools and temporary timbers may be borrowed. A roof over the ways is really well worth the trouble of rigging, and can be very light, covered with tarpaulins and guyed like a tent.

SHELVES AND PLANKING

Counter Construction

The horn timbers and midrib, shown in Fig. 56, can be attached to the sternpost before erection, and the whole of the counter and after deck is supported by them. The rudder trunk is lined with $1\frac{1}{8}$-in. pine battens, arranged with a taper down from the deck partners to the horn timbers, so that they can be wedged tightly into place (with a luting of white lead), after the rudder has been hung.

The counter frames are built up as usual with double futtocks, butted to the horn timbers, and secured by floors. The shelf does not go right to the stern frame, but supplementary clamps are often fitted across the aftermost four or five frames to help support the stern. Such clamps are not necessary in this short overhang.

Our counter is slightly sawn-off, with a transom below the taffrail, and knees are bolted to the stern covering board and beams to support the taffrail and transom planking. A large knee at each side is curved to the tumblehome and has rabbets to receive the bulwark and transom plankings.

Such a stern is very attractive, with the transom planking laid athwartships, but radiating in plan, from flat at the stern frame, through a slight curve at the covering board, to a very full ellipse at the taffrail.

Shelves and Planking

The shelves can be fitted either before or after planking, but before is better, because the hull is made slightly more rigid for planking, and the bolts which run right through from the sheer and/or binding strakes to the shelves can be driven as soon as these strakes are fitted. The shelf is very stout, of 5-in. square section oak, with not more than two scarphs per side, and fastened to the

CONSTRUCTION OF A LARGER HULL

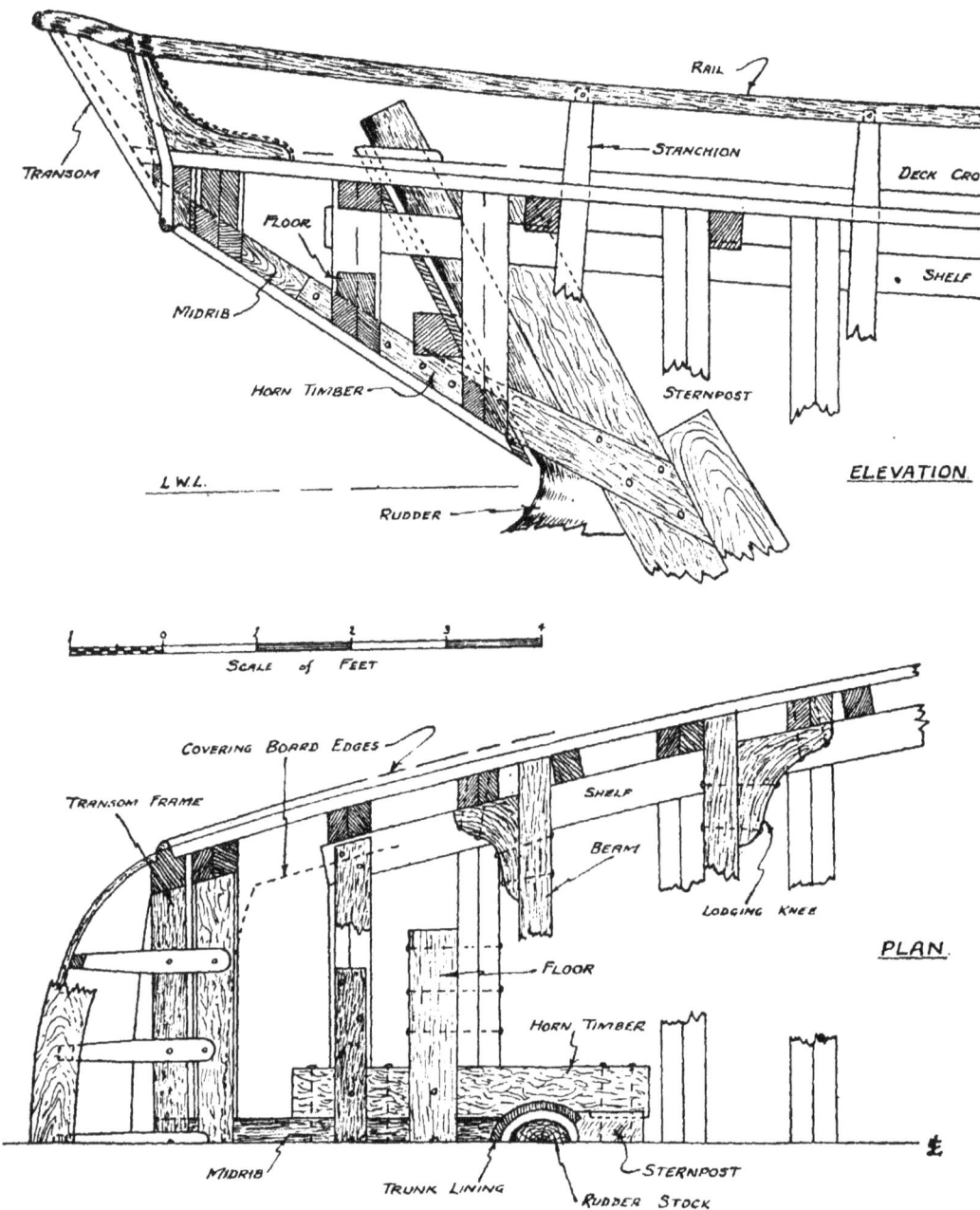

FIG. 56.—Construction of the counter.

SHELVES AND PLANKING

timbers and planking with two staggered bolts at each frame.

To determine the length of a scarph, give it an angle of 13 deg. to the side of the timber as a safe minimum. More is better. Each piece will require prolonged steaming, and then clamping to a form to get the approximate curve before trying it in the boat.

The fore-and-aft curve must be picked up from the

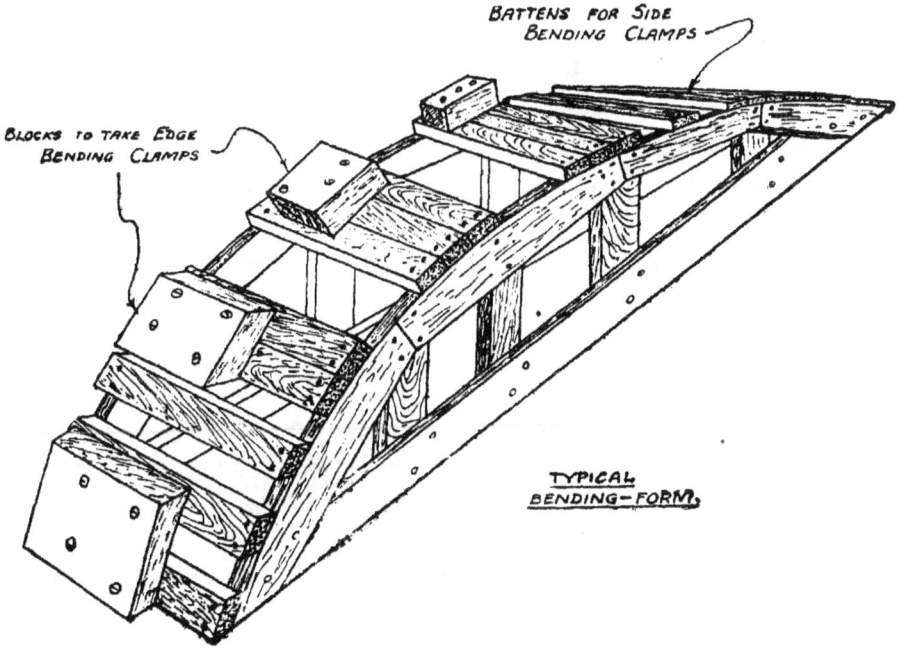

FIG. 57.—A typical bending-form for the shelves, etc.

mould loft floor, and the sheer curve must also be applied, so that the form should appear as in Fig. 57, needing two separate sets of clamps to hold the shelf. Very often, only one set of clamps can be used at a time, so that after setting for a day, the piece must be re-steamed to clamp the other way. Make the jig of smaller radius than is required, as the timber will straighten out slightly after removal.

CONSTRUCTION OF A LARGER HULL

Planking is very similar to that described in Chapter VIII, except that here there is a strong reverse curve to the bilge aft. The order of fitting strakes is the same, and they are 1¾ in. thick, except the binding strake, which is 2¼ in., more for appearance than needed strength. The planks follow the diagonal lines closely, and the arrangement can be sketched out quite accurately on a body plan drawing.

It will be found that the garboards are very wide aft, needing especially good stock to cut them, and that is the reason why so many working craft are seen with one or two stealers worked in right aft. Actually they are no source of weakness here, right against the deadwood, but look bad in yacht practice, unless the hull is copper sheathed !

The stock for each plank must be carefully chosen, so that the ones with greatest bend are cut from the best vertical grain (slash-sawn) stuff, and the wide strakes have no sap wood in them. If oak is used, most of the stock will have some sort of general curve in its length before cutting ; this can be made use of, especially for the sheer, garboard and adjacent strakes. A plank which has upward curvature thus, is said to sny ; or, with downward curvature, to hang. Therefore, it is not only because of the better strength and toughness that so many craft have oak in the regions of the sheer and garboards, and the remainder of the skin of straight stock, such as pine or teak !

Readers who have perused that yacht-builders' pet book, *The Southseaman*, by Weston Martyr, will remember how much importance the Nova Scotian shipwrights attached to the picking of a piece of timber for planking. Due to reverse curve in the bilge, the plank edges have pronounced bevel in many places. This bevel need be applied only to the fitted edge of the plank, the other

JOINTS AND SCARPHS IN PLANKS

edge being left square for the next strake to bear all the bevel. " Soling " and thicker planking may be necessary—see Chapter VIII.

Joints and Scarphs in Planks

Although butt joints in planks are common practice in all types of craft, and are quite strong if properly fastened to a butt block (which may have to be curved slightly to fit snugly to the inside of the planking), recent tendencies with high-speed boats, built to Admiralty specifications, have introduced scarphed plank joints, which may also have a block behind, but which more frequently have not. Instead, the scarph is very often situated directly across a frame, and not exactly between two frames as the general practice with butt blocks necessitates.

These plank scarphs must be cut as trailing scarphs, so that should the joint start, and spring open, the rush of water passing the hull will tend to keep the plank in place; not to rip it off. A chine rubbing strake has been known to tear right off in this manner on a motor launch.

There is one other important consideration where these particular scarphs are concerned. They must be well made and very strong, as such thin members are joined. In most instances where scarphs are used, such as in keels, stems, stringers and rails, the timber is amply strong, and the joints are covered by other members, so that almost any design of straight, wedged, or tabled scarph may be used, providing its overall length is greater than three times the depth of the timber.

Theoretically, the strongest possible scarph is one made as in Fig. 58 (A) with a straight cut, but this type possesses the big drawback of curling edges with consequent splitting, unless both sides can be protected in some way. Furthermore, as no shear strain should be imparted to

CONSTRUCTION OF A LARGER HULL

the fastenings, an important scarph should be tabled, or wedged, so that it will not slip even with the fastenings removed. For plank scarphs, the roved copper nail fastenings are so numerous, and closely spaced, that no locking device is necessary, but to maintain maximum strength, the length should be about four times the plank

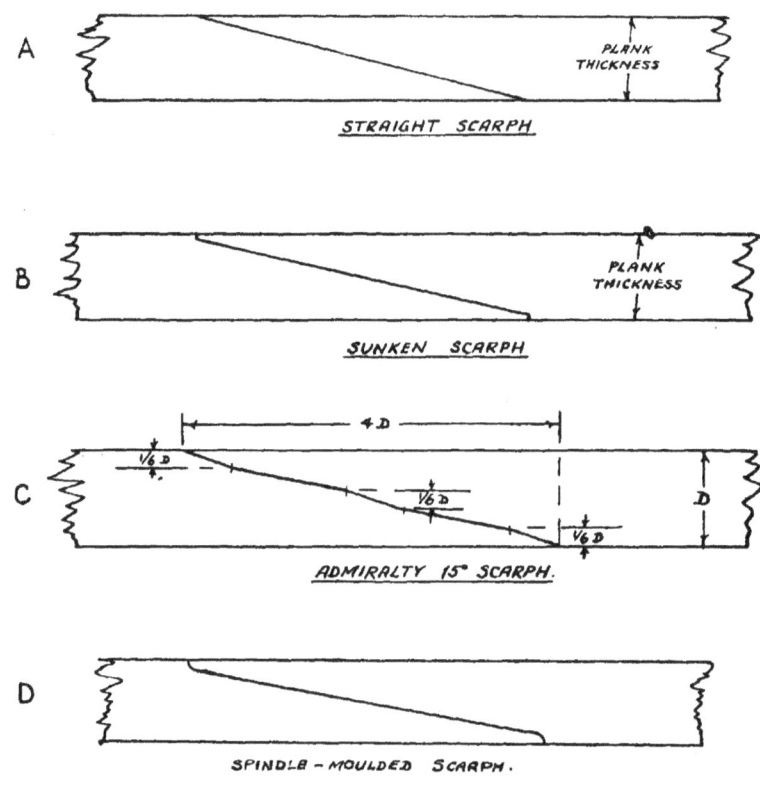

FIG. 58.—Types of planking scarphs.

thickness, preferably more, and just a slight notching of the tips effected, to avoid curling edges, as shown in Fig. 58 (B).

The least acute angle required in normal hardwood to avoid a curling edge is 15 deg., so the Admiralty pattern scarph is designed as in Fig. 58 (C). The purpose of the flat in the centre is somewhat obscure.

PLANK FASTENINGS

Very strong and efficient scarphs can be made on small members with the aid of a spindle machine, with a result as shown in Fig. 58 (*D*). This avoids sharp corners from which cracks may start when a stress is applied.

Plank Fastenings

Plank fastenings should all be of through bolts or clench nails; dump fastenings cannot hold properly if of copper, and are false economy in an iron-fastened boat. Trenails are rarely used in small boats, but two $\frac{3}{4}$-in. trenails could be used on every alternate frame in our example. Besides cheapness, resilience and long life, they have no advantages over metal, though one could count the elimination of countersinks and stopping as a merit.

Trenails should be of oak, with the grain straight across in section and running end to end. Each end is split centrally *across* the grain, the trenail is hammered into a very tight hole (often with a metal cap to prevent splitting), and a narrow oak wedge, in our case $\frac{3}{4}$ in. wide by $\frac{1}{8}$ in. thick by $1\frac{1}{4}$ in. long, is hammered into each end-split. A dump trenail can be made by fitting a wedge loosely into the far end before driving it into the blind hole. On reaching the end of the hole, this wedge will then be driven home, and the trenail firmly held, while the outer end is wedged in the customary manner. Trenails must be driven with great care, and they are inclined to involve more work than similar metal fastenings. The amateur builder should try fitting a few into scrap timber before he decides to adopt them.

With trenails, one bronze through-bolt should be used at every alternate frame, and in all butts (which terminate at the centre of a frame, with each plank end having two bolts into a futtock). The hood ends, especially at the stem, must be very well fastened, or a collision may start them all. This is best accomplished, at the stem, by

CONSTRUCTION OF A LARGER HULL

putting two bronze bolts right through between each pair of planks. Copper clench bolts could be used for the stern ends.

The stringers must be fitted at the same time as the planks they come opposite to, and the whole through-fastened. The shelf will be similarly fastened and the breasthook put in after the sheer strakes are fitted. If a ceiling (or lining) is fitted, it is normally dump fastened to the frames, but a cleaner and better ventilated arrangement results if the ceiling is omitted, and the timbers and planking nicely finished for painting and enamelling above the stringers. The chain plates should be bolted on before any ceiling or interior fitment goes in.

Deck Structure

Deck beams can replace the cross spalls on the frames, and are notched and fastened to the shelves as shown in Fig. 59, with lodging and hanging knees fixed as drawn. Lodging knees should be used at every beam, preferably one each side, but hanging knees need only be fitted in way of the mast and chain plates. The beams are $4\frac{1}{2}$ in. deep by 5 in. wide, except for the two bearing the mast partners, which are $4\frac{1}{2}$ in. deep by 7 in. wide.

Carlines notch into the beams, but are straightforward carpentry, and all this structure should be of English oak. After smoothing off the upper surfaces of the beams, knees and sheer planks, the covering board can be fitted. This is 9 in. wide, $1\frac{7}{8}$ in. thick, sawn from oak having a natural sweep; in not more than three sections.

For this, and all decking, galvanized iron dump nails (the square boat type) will be used, as there is no fear of electrolytic action or weep marks, and they hold well. The square holes for the bulwark stanchions may be cut in the covering board with a taper for caulking before the latter is fastened, and the stanchions can then be made

DECK STRUCTURE

a tight fit and bolted to two or three strakes of planking below the sheer, and between timbers. A tenon on the top of each stanchion is morticed into the rail.

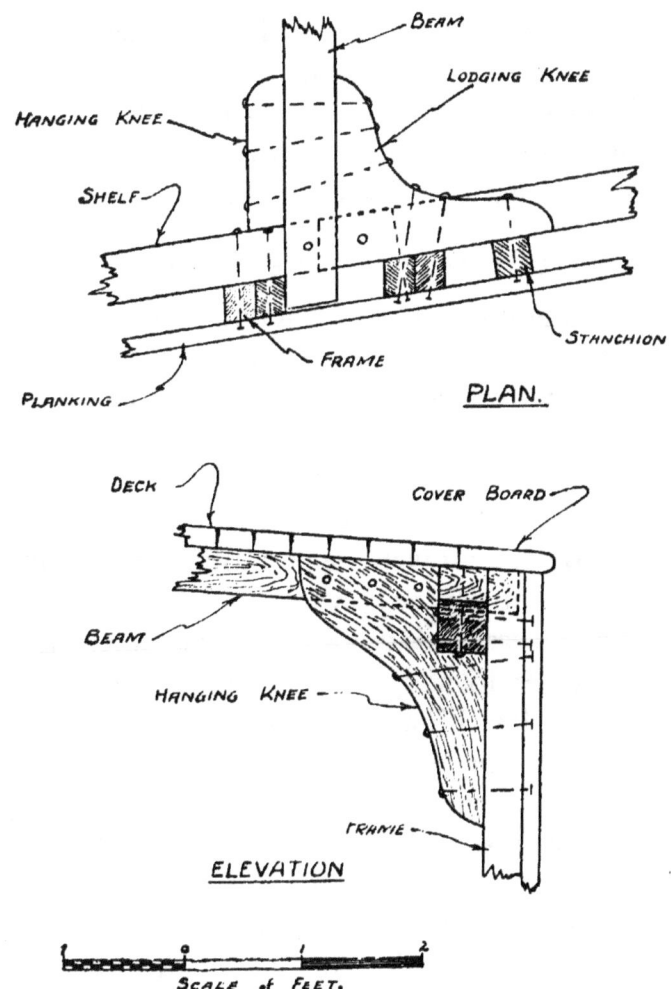

FIG. 59.—Deck and deck beam fastenings.

The deck planks, of pine or teak, are $1\frac{7}{8}$ in. thick by 4 in. wide (with the usual caulking chamfer) laid parallel to the king plank. The edge fitting must be perfect, to prevent leaks, and no working of the seams will ever occur if $\frac{1}{2}$-in. oak dowels are fitted between the edges,

CONSTRUCTION OF A LARGER HULL

every 8 in. or so. Clamp tightly into place whilst nailing, punching into the holes previously bored for the heads, which are then plugged with dowels set in varnish.

Planks are usually laid from the centre of the deck towards the covering board, and the holes should be drilled with a slight angle towards the king plank. Choose a drill of a size to ensure a driving fit in the plank, and use a smaller drill for the remainder of the hole into the deck beam, so that a really tight drive results.

BUILDING WITH STRAIGHT-GRAINED TIMBER

Now the method of construction just described is similar to that used for the largest wooden vessels, and differs but slightly from that used several centuries ago. True, given the same materials, there is no other sound way of using them, but prohibit the use of natural crooks (a situation which did, in fact, arise during the war, when many big wooden craft were being built for the Admiralty) and an almost entirely new system is bound to evolve.

Even when seasoned crooks are available, there are many advantages in adopting a design which uses only straight timber, for although under these conditions there must necessarily be a higher percentage of waste, it is found in practice that the overall cost is lower, and a lighter hull results. Such a system is obviously of interest to the amateur builder, but as it has not been universally adopted by naval architects, it would probably mean altering the designer's specification and constructional drawings in order to comply with the following descriptions.

To introduce every aspect of " crook-less " building, we will now see how it would be applied to the 48-footer hull just described.

The keel assembly would appear as in Fig. 60 (a). Quite conventional except for the stem, and for the hog,

BUILDING WITH STRAIGHT-GRAINED TIMBER

which is unusual in so large a boat. At this stage, the keel scarphs must be fastened as scantily as possible, to allow for the many subsequent keel bolts. It will be apparent from the section shown in Fig. 60 (*b*) that the cutting of the planking rabbet is simplified considerably by the adoption of a hog, as it then devolves into the making of two chamfers instead of a chiselled hollow and the hog also provides a much wider seating for the lower

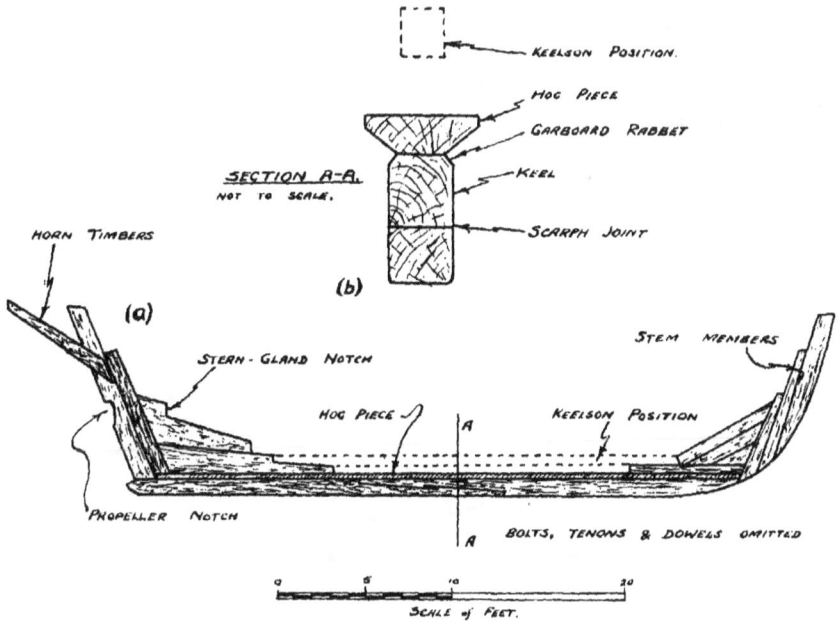

Fig. 60.—The 48-footer built with straight-grained timber.

edge of the garboard strake and its fastenings. A ballast keel could, of course, be readily fitted if desired.

Next we come to the frames, which are to be spaced exactly the same as for the old type of construction, but which are made of double futtocks each $1\frac{3}{4}$ in. thick, giving a complete frame thickness of $3\frac{1}{2}$ in.—or only slightly more than half the thickness of the old crook futtock frames. The reason why this big reduction in weight and material is possible must now be explained.

CONSTRUCTION OF A LARGER HULL

First of all, no through fastenings from the planking or ceiling are to be used : screw fastenings are adopted for all these. Therefore, there is no weakening of the frame timbers by the boring of bolt holes, or by the driving of long clenched spikes. Secondly, the futtocks are not merely butted end-to-end and then trenailed or spiked to their partners; they are scarph-jointed, as shown in Fig. 61, and then fixed together with heavy copper nails

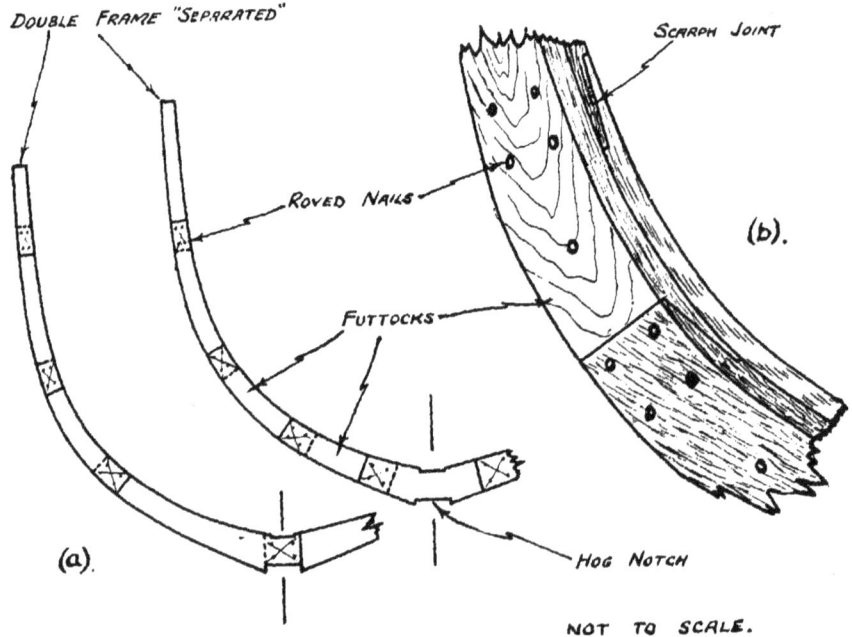

FIG. 61.—Details of " crook-less " frames.

closed over roves at close intervals. Thirdly, the moulded width of these frames is correspondingly greater, adding more to strength than a big increase in sided thickness would do, and allowing long screws from the planking and ceiling to be used without undue weakening of the timber. As no crooks are used, the futtocks must be short, so that scarphs are numerous, but as these joints are well fitted and thoroughly fastened with roved nails

BUILDING WITH STRAIGHT-GRAINED TIMBER

passing right through to the partner, the strength is reduced but very little from that of the solid wood. Fig. 61 (*a*) shows the two halves of a typical frame separated to reveal the positions of each scarph joint. Fig. 61 (*b*) gives the details of the futtock fastenings.

The notch shown in the bottom futtocks brings us to the keel fastenings at each frame, which are arranged as

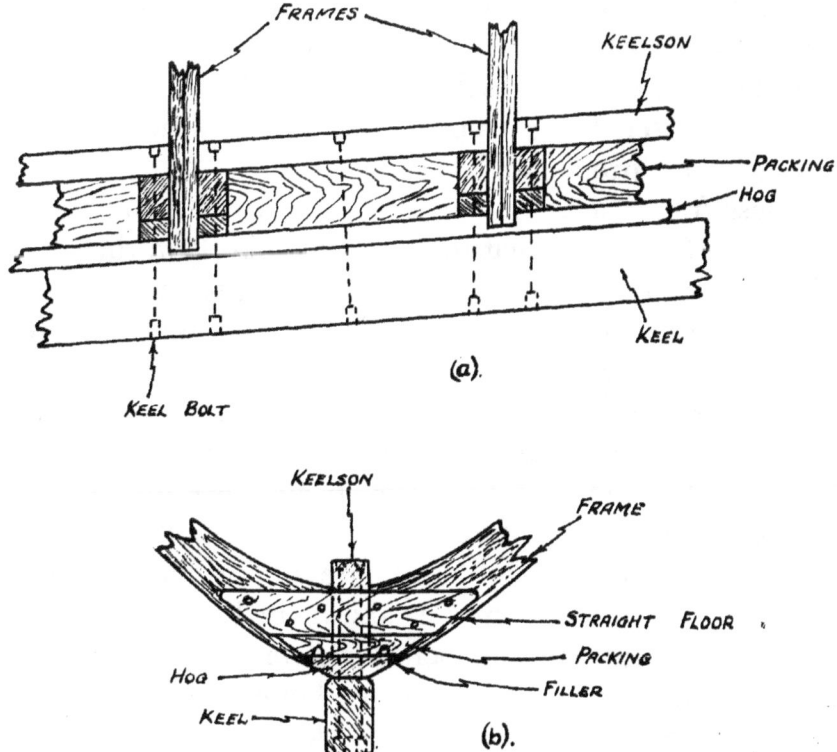

FIG. 62.—Straight-grained frames and floors.

in Fig. 62. The notch fits over the hog. Two stout pieces of timber are bolted to the sides of the frames to take the keel bolts (without necessitating the boring of holes through the futtocks, which would weaken the frames). The keelson lies on top in its usual position, and two staggered keel bolts at each frame pass right through, as shown dotted in the drawings. In the spaces between frames, timber blocks are fitted to fill the gap between the keelson and the hog, and a bolt can be driven at the centre of each of these blocks for extra strength (though it should be a light bolt to avoid weakening the timber). An advantage of this block is that it divides the bilge into two separate halves, so that by providing two separate intakes for the bilge pump, one each side of the block, the bilge water may be completely drained at any angle of heel. Filler pieces are fitted alongside the hog, between each frame, to mitigate the likelihood of water lying trapped in spaces which would otherwise be prone to rot.

These frames are generally moulded wider than similar oak crook frames, so that the excessive bevels in the ends of a vessel can be applied without resorting to cant frames. Where the frames sit upon the deadwoods, fore and aft, it becomes impracticable to bolt them as thoroughly as where they are flat on the hog. Therefore, to equalize the strength in these quarters, steel plates are bolted between the frames. The deck beam-to-frame attachments are unusual, and extremely neat, as well as strong and easy to build. The beams are simply half-notched into the tops of the frames as shown in Fig. 63 (a), and the whole joint is sandwiched between two straight-grained hanging knees which are bolted to each other through the centre pieces. The deck beams are built up with long futtocks, though they could quite well be solid if the slightly curved crooks were available.

In vessels with high bulwarks, the frames are taken on up through the covering board to form the rail stanchions, but in a small yacht it would be better to fit separate stanchions, fastened to the planking in the conventional manner.

A shelf and clamp can be fitted, notched into the ends of the hanging knees, but a neater and equally strong

BUILDING WITH STRAIGHT-GRAINED TIMBER

method is as shown in Fig. 63 (*b*), where a very stout sheer strake makes up for the loss of an internal shelf, while it also serves as a valuable rubbing strake to protect the topsides.

Planking is double skin laid fore-and-aft. Using 1-in. thick skins in the 48-footer, with a 2¾-in. solid or double sheer strake, the result will be far stronger than the corresponding single skin job with caulked seams. Further-

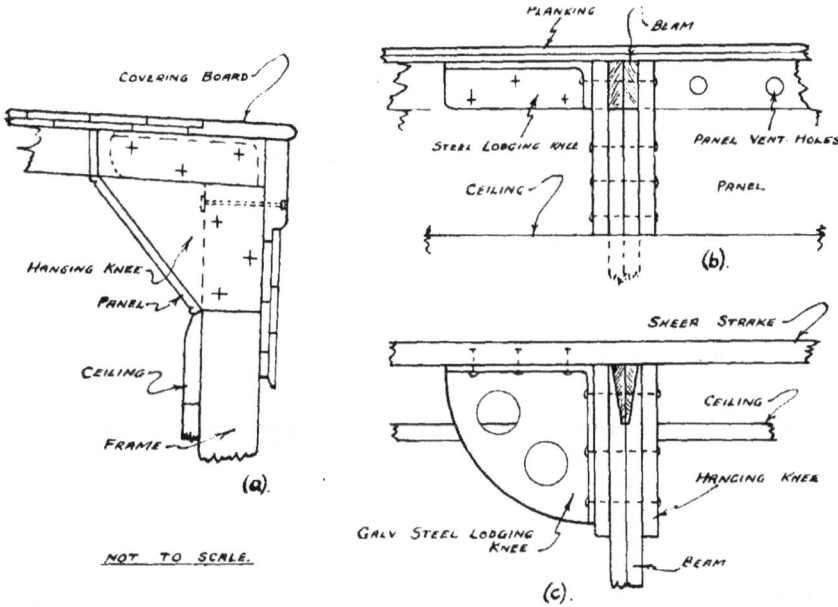

FIG. 63.—Knees and double planking.

more, leaks will be eliminated, and no caulking or re-caulking will ever be necessary. This method of planking has never been very popular because of its expense, but it has been perfected by the Ahmadi Shipyard in Bombay. They agree that it is more costly than any other conventional system, but for the extremes of climate experienced in that locality it is ideal, and cannot be bettered for any other region. This fore-and-aft, or half-lapped double skin planking has one very

CONSTRUCTION OF A LARGER HULL

great advantage over double diagonal—it is far easier to repair strongly. A really first-class hull repair is almost impossible with double diagonal, and necessitates endless stripping of the outer skin, to reach the affected inner strakes. The simpler bending of 1-in. timber makes the system advantageous for amateurs, who rarely have the labour to handle heavy strakes, and for the same reason, rift-sawn planking can be used, giving a cleaner finish, and less liability to cracking and warping.

A good hardwood that is little affected by weather conditions must be used. Teak is ideal. . . . The deck is laid in a similar manner, and can therefore be really tight. Lodging knees of welded hot-galvanized steel are bolted into the corner on one side of each frame, as shown in Fig. 63. The spaces above the ceiling are then covered with removable panelling, the sections between the deck beams being drilled with large holes to provide ventilation down to the bilge.

Many modifications of this system are conceivable, and certain parts could use crooks, where it might prove cheaper, or easier. More labour is involved on the whole. The method has been adopted by the Admiralty, and is fully approved by Lloyds' surveyors, but care must be taken in getting the sizes of scantlings and fastenings exactly right, so that no member is superfluously strong, but so that strength is evenly distributed throughout the structure. That is the secret of good design.

PART THREE

MISCELLANEOUS

CHAPTER XI
VENTILATION AND CABIN CARPENTRY

VENTILATION

MANY good designs for small yachts do not show any specific devices to effect adequate ventilation, or else arbitrary methods are merely suggested. Ventilation is extremely important, even in the smallest craft, and it cannot be overdone. The chief methods employed necessitate the installation of several intake spray-proof ventilators in the fore-deck, to create a draught through the bilge and accommodation, which is extracted by exhaust vents set in the after deck. If the bilge is thus well aired, the cabins will automatically smell sweet, and the timber will be kept dry and free from rot. Engine compartments should be separately ventilated.

Many excellent types of spray-proof vents are available, but none are more efficient than the old Dorade vent, which can be built up with wood, and is best set in the fore side of a hatch, skylight, or cabin top, where the box will not prove unsightly. The general arrangement is shown in Fig. 64, and the amateur boat builder can modify this structure to suit his own vessel. A draught control, adjustable from inside, is almost essential on any intake vent, and the inlet can be at deck level unless a particular vent is made to serve the bilge only, in which case a tube should be led from the cowl to below the cabin sole.

Extractor vents are commonly made by simply revers-

VENTILATION AND CABIN CARPENTRY

ing a cowl, so that it points down-wind. A more efficient type may be constructed quite easily from sheet metal, to work on the Venturi principle, as sketched. An extractor vent need not have a draught control, but a normal Venturi exhauster is not spray-proof, so that a shutter should be fitted below, to be closed in very heavy weather.

All spaces below berths and inside lockers should be

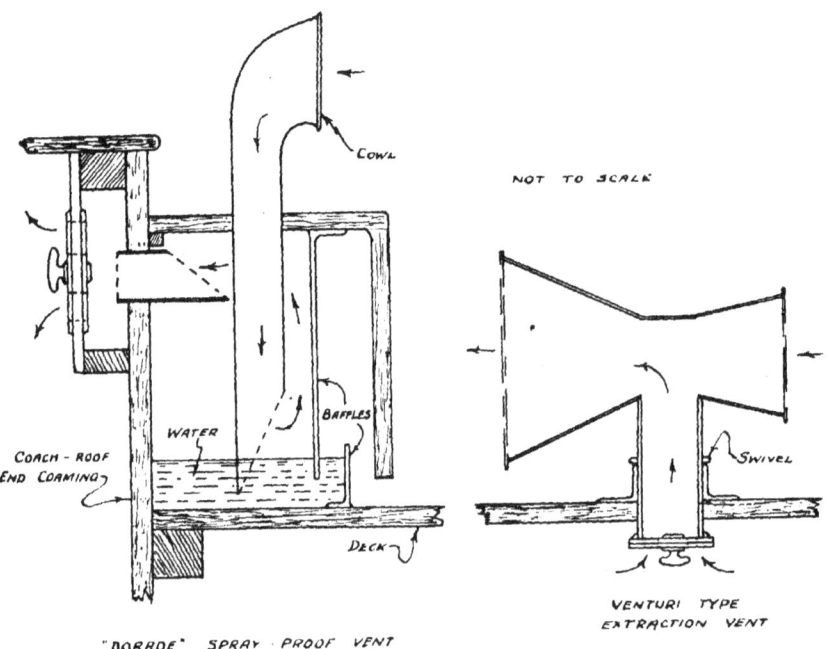

FIG. 64.—How to exclude water.

ventilated by means of louvres or plain holes, and even cabin doors should have their upper panels louvred, or fitted with some kind of adjustable vent. No compartment should be completely sealed.

A ceiling should be dispensed with in vessels smaller than about 15 tons T.M. and replaced by stout battens, thus ensuring perfect ventilation to the bilge. It is better still to finish the frames and planking nicely and enamel them, but battens are used instead because of

CABIN CARPENTRY

the difficulty of finishing the structural parts neatly. When battens are fitted the planking and frames should be creosoted black before fitting the battens and, if the latter are then painted and enamelled white, one cannot see beyond them and a pleasing effect is produced. Berths should also be formed of battens, fixed fore-and-aft to slightly hollowed bearers.

CABIN CARPENTRY

There is nothing difficult about the many other tasks which remain to be done after the building of a hull, to transfer the shell into, say, a cabin boat. The processes which are essentially boat building have already been treated at length; the cabin tops, bulkheads, hatches, lockers, and berths are simple carpentry, and should prove very enjoyable to the builder, for once his hull is fashioned he can fit it out almost exactly as he wishes.

The important parts are those which exclude the external elements: the cabin top, decks, skylights, hatches, and scuttles. The timber for the coach roof must be very stout, and strongly fastened, so that no weathering effects, or forces produced by green seas can move anything to cause a leak. Skylights are especially important, as they are known throughout the yachting world as sources of almost incurable leaks. Therefore, skylights should be especially strongly constructed, preferably of teak (that will never warp or contract) or Honduras mahogany. Armourplate glass (as made by Pilkingtons) is well worth its extra cost, and need not be provided with the conventional ugly guarding bars.

The entire weakness of the majority of skylights lies in the hinging of the tops. Gutters can be fitted below the hinges, and all seams lined with rubber, but still such a skylight will leak, so, therefore, the amateur is advised to adopt a skylight without hinges of any kind. This

VENTILATION AND CABIN CARPENTRY

type, sketched in Fig. 65, is made so that the entire structure above deck can be removed in harbour, or in fine weather at sea, while normally at sea the whole skylight remains battened down, with a spray-proof

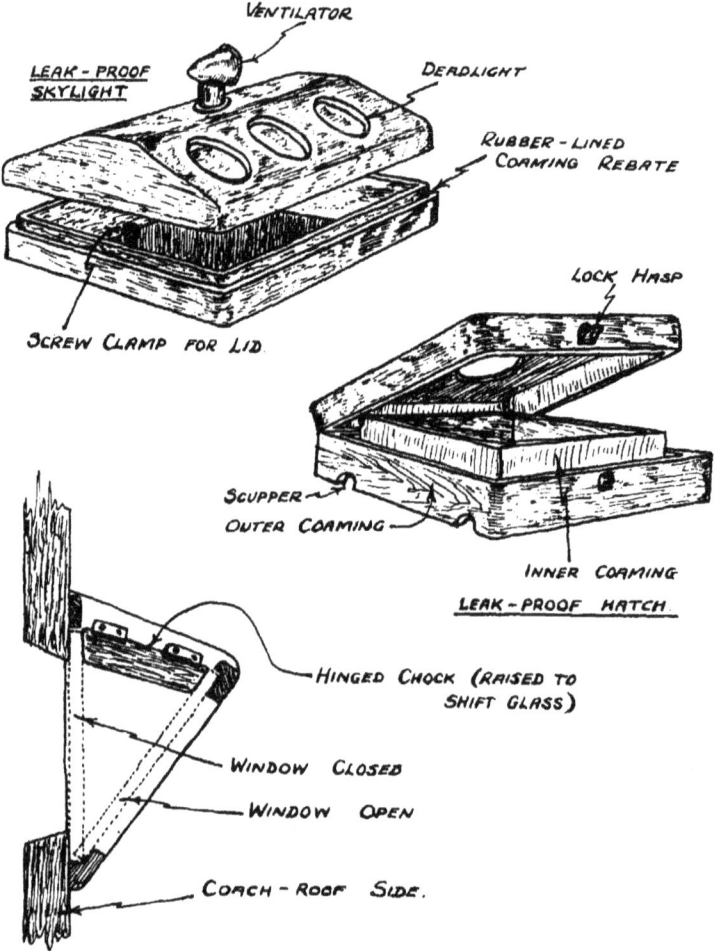

FIG. 65.—Ventilators.

ventilator set in in the top to keep the cabin airy at times when the lee flap of a conventional skylight might be opened; the internal draught control can be completely closed in rough weather. The construction is further simplified and strengthened by fitting round deadlights

CABIN CARPENTRY

instead of normal square glass windows. All glass should be set in Bostik, or some similar plastic rubber compound.

Such a skylight is essentially a hatch, in fact the top could be hinged along one side without much increase in the likelihood of a leak. It proves ideal in really hot weather, when the lid can be removed and a wind scoop, or wind-sail of canvas fitted to catch the breeze.

The best type of water-tight fore-hatch is that sketched alongside the skylight. The lid closes on to a false coaming, so that any water leaking through the lid joint is trapped in the space between the coamings and expelled through the scuppers provided.

Where rectangular windows are desired in coach roofs or companion hatches, they may either be the conventional brass-framed fittings, or else the very simple and cheap device sketched herewith. The plate of glass (which should be of the Armourplate variety) can be locked in two positions by means of the hinged wooden chock—either tightly closed, or fully open.

CHAPTER XII

SHEATHING, BALLAST, CAULKING AND STOPPING

Copper Sheathing

TO prepare the bottom, first make sure that all the caulking and stopping is perfect. Then apply hot coal tar, and stick sheets of well-tarred stiff brown paper over the whole of the planking that is to be sheathed, i.e. everywhere below a line 6 in. above the load waterline.

The copper should be 22 gauge (Imperial Standard Wire Gauge), in sheets 4 ft. by 14 in.; except for very small boats, where 26 gauge copper may be used, in much smaller sheets.

Lay the sheets from waterline downwards, and from stern to bow, overlapping them 1 in., and nailing around $\frac{1}{2}$ in. from the edges with copper sheathing nails, at intervals of 1 in. Also provide three rows of nails across the remainder of each sheet, these nails at intervals of 4 in. Sheathing nails have a serrated shank and should be 1 in. in length except for the smaller boats, for which the length should be $\frac{1}{4}$ in. less than the plank thickness.

The rudder may have to be unshipped, and sheathed separately, but for a vessel of over 8-tons T.M., commence at the stern and lay two or three rows of sheets the length of the hull, before starting at the garboards and working upwards, to leave a " shutter strake " of sheets for completion. Make sure every sheet is firmly bent to shape, by driving the centre row of 4-in.-spaced nails first, and hammering the sheets to the curve of the hull with a wooden block and mallet, or coppering hammer (which has a slightly curved surface). Separate cut sheets must be lapped round the stem, sternpost, and

BALLAST

under the keel, and for very severe bends these small sheets may require annealing to render them more pliable. Heat them to redness in a furnace, or improvised wood fire, and then plunge rapidly into cold water to effect annealing.

Do not sheath over a lead or iron keel, but fit a strake of roofing felt for 6 in. from the top edge of the keel upwards. The lower edge of the sheathing should be kept about $\frac{1}{2}$ in. clear of the other metal to minimize galvanic action.

Before commencement of sheathing as many as possible of the nail holes should be drilled or punched. All the central rows of holes can be made, and these should, of course, be staggered. Usually only two of the rows around the edges are drilled beforehand, i.e. those holes which are uppermost at the overlaps. A wooden template had best be made to ensure even spacing of the holes. The holes are usually made with a point punch, and hammer, but an electric hand drilling machine will enable a number of sheets to be done simultaneously and produce cleaner holes. A special board is usually made up, with strips of wood forming a frame which just takes the sheets, and positions them whilst punching or drilling. When laying, " point punch " through the already drilled holes to puncture the lower sheet.

The same rules apply for sheathing in zinc, but in this case smaller sheets are advantageous, for the metal is less ductile and will not shape so easily to the curves of the hull.

Ballast

The material and quantity of ballast is invariably specified on a proper design, but even then certain troubles may occur. For instance, most old designs tend to specify lead for both internal and external ballast, for that

SHEATHING, BALLAST, CAULKING, STOPPING

metal used to be far cheaper than it is these days. Lead has few advantages over iron for this purpose, except, of course, that it occupies less space, due to its greater density, and it does not corrode so easily as iron.

The ratio of the density of iron to that of lead is about 1 to 1·6, depending on the grade of iron, so it is quite a simple matter to calculate the size of an iron keel to replace a lead one shown on a design, and to find if there is sufficient space below the cabin sole to house all the inside ballast required, when changing it from lead to iron. Providing the centre of gravity of the iron keel is kept to conform to that of the old lead one, a little trimming of internal pigs should correct any other change of trim.

Where space for inside ballast is restricted, this may have to be specially cast, to fit the curve of the bilge, but a cheaper job results if there is sufficient room for all the internal ballast to be composed of pigs of a standard size. These may be about 2 ft. long, by 6 in. square in section, and cast with a hole through the centre to facilitate lifting. Lead ballast need not be treated, but iron should be coated with two coats of either black varnish or red oxide paint. This surface will last a very long time if the pigs are cast with well rounded corners, and rigidly stowed.

In fitting internal ballast, see that none of it is allowed to rest upon the planking. Fit special wooden chocks between the frames to support it, but if the hull has a ceiling right down to the keelson, the best place for the ballast is on top of this ceiling, and below the cabin sole, where it will be well away from any normal bilge-water. Avoid using concrete in the bilges of a small boat, which would not be allowed in any case if the building was to be inspected periodically by a Lloyds' surveyor.

CAULKING AND STOPPING

Caulking and Stopping

The methods of finishing, caulking and stopping for a small carvel-built hull were described in Chapter IV. For a larger craft the finishing process would be similar and during its progress an examination can proceed to ensure that all fastenings have been driven and that no timber has split badly since fitting. Caulking and stopping, however, will require a slightly different technique.

The seams must be cleaned out carefully, to rid them of the fine shavings and dust that will have gathered there. It is usually better to scrape out about ten seams at a time rather than each one separately, just prior to caulking. The special caulker's " clearing " iron can be used, although a stiff seam brush or an improvised tool made from a piece of tapered and hooked 12-S.W.G. steel wire set in a wooden handle serve the purpose just as well.

Caulking cotton must be used for seams in planking under 1 in. thickness, though it can be used up to 2 in. However, oakum is more often used in the thicker planking, for it is much cheaper and lasts longer if of good quality. Cotton is bought in bundles, or hanks, composed of strands of varying thickness as a rule, so that the best strands for a particular width of seam can be chosen. Oakum comes in bales, from which wads of it can be torn, and a few of the yarns teased out in as long a length as possible to be rolled up tightly between the hands, or between one hand and the thigh. The prepared strand of oakum, or the chosen strand of cotton, should be about the same thickness as the width of the seam opening.

There need be no rigid rules as to how caulking should be driven, but the recognized method is to hold the " making " iron in the left hand, and work from left to right along a seam. The iron is held exactly the opposite way to a chisel, and rests across the palm while the back of the hand faces downwards (when caulking the vertical side of a hull), so that the iron can be grasped loosely by the curled-over fingers. The strand of cotton or oakum can be fed between the first two fingers, while the mallet is held in the right hand. Lay the cotton in a series of overlapping loops, each about 2 in. long, but, unless the seam is very deep, drive oakum as a straight strand without loops.

The amateur will find the caulking mallet somewhat heavy and awkward to use at first, although it has the solid driving power that is required. However, a good compromise is to use a normal light carpenter's mallet for placing the caulking, and then go over the seam again afterwards with a proper mallet and caulking iron to drive more effectively, and to make sure that the cotton is pinched in the bottom of the seam, and is $\frac{1}{16}$ in. or so below the plank surface, to leave room for the stopping, or paying.

If a seam is too narrow, it may be prised open slightly by means of a " dumb " iron, having a tapered blade, but if the seam is very wide, or open right through, the caulking can be driven with waterproof glue or paint to ensure it stays in position. If the seam is too wide even for this process, the faulty plank had better be replaced while the opportunity avails itself, but sometimes a spline of timber can be driven as a long wedge into the wide seam, together with a strand of oakum and some paint. The spline can then be nailed to the edge of one of the planks. Painting the seams with white lead just before the caulking is driven is actually very good practice, as the cotton or oakum is thus anchored more securely, and has a longer life.

Finally, the seams may be payed with one of numerous

CAULKING AND STOPPING

good preparations. A tarred hull, or almost any bottom seams can be payed with a mixture of hot pitch and tar, applied while of the consistency of thick paint by means of a seam brush. This stopping adheres to the threads of oakum as well as to the wood. It remains permanently slightly elastic, and is very cheap. Many patent marine glues are obtainable for this purpose, often superior to any home-made preparation and worth their extra expense, especially the topside variety that can be painted without showing through. Red lead putty is a good substitute for any commercial stopping, but it is too troublesome to use for the seams of a big vessel unless a " seam-gun " can be utilized to force a stream of putty over the caulking. More paint should be applied with a seam brush just prior to paying with putty, as this will check the absorption of the linseed oil in the putty into the plank edges, with consequent cracking and loosening. Unless fastening countersink holes have been dowelled over, they must be filled with putty. Substitutes, such as cement mixed with marine glue, or sawdust mixed with varnish or gold-size are often adopted on big craft.

Deck seams may be treated in exactly the same manner, but work is facilitated considerably by the horizontal position.

CHAPTER XIII
A HALF-MODEL

VERY few yachtsmen can visualize the exact appearance of a hull from looking at the lines drawings. Most keen cruising men have an eye for a good hull when seen on the hard, and when planning a new boat or trying to decide the best of several designs, the only sure and practical way is to make a half-model out of wood to see if the one which appears ideal on paper would give the same effect when actually viewed on the hard. The reader can then examine his future boat before she ever takes to the stocks, and he will know that the model will be reproduced exactly in the real hull, because both will be made from the same set of accurate drawings. The following will show how simple the making of a model can be, and may make all the difference between success and failure in the completed yacht.

BUILDING THE MODEL

The model shown in the accompanying photographs was made from the lines drawings of the pilot-cutter-type yacht discussed in Chapters IX and X. About thirty man-hours were spent in making it, most of these being absorbed by the finishing processes. If the builder is keen on modelling he could, of course, make a complete model and rig her, but for the purposes of testing the design a half-model is adequate and even that can be a respectable ornament.

If the model is made to the same scale as the plans it will be just the right size, and building it will be simple. No lines drawings for a cruising yacht should be smaller in scale than half an inch to the foot, and neither should

Plate 16.—Views of a " bread and butter " model during assembly.

A HALF-MODEL

a half-model. This means that the cutter illustrated is about 2 ft. long and 6 in. deep.

Such a model could be made from a solid block of wood, and carved out with gouges and chisels until templates conforming to the station lines of the body plan fit exactly to the curves at their respective places along the hull. But a far simpler way is to build up the model from a number of laminæ, or layers of wood screwed together, as shown in Fig. 66, so that a mistake will not necessitate the scrapping of the entire job. Laminated construction further simplifies the task, because the joins formed where adjacent laminæ meet can be made to correspond to the waterlines on the profile and body plans, and can thus be sawn approximately to shape before the pieces are screwed together. As well as these laminæ forming the hull, a backing piece must be made to the shape of the profile plan to form the parts of the stem, keel and sternpost which protrude beyond the planking rabbet. This backing piece should, strictly, be of a thickness equal to half the thickness of the stem and keel, but a more realistic effect and a stronger job will result if this board is made the full thickness of the keel. The

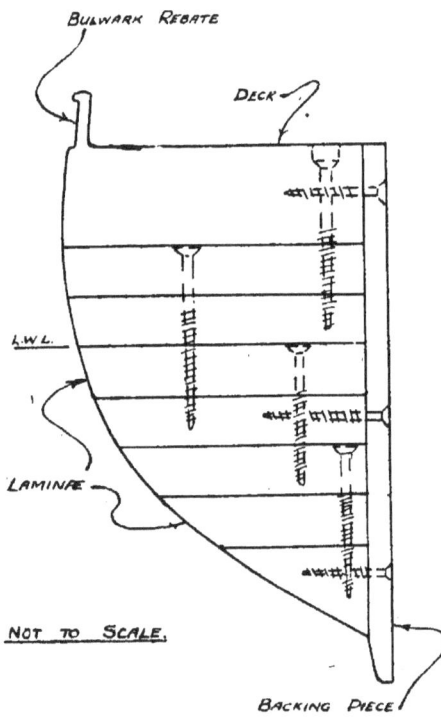

FIG. 66.—The plank joints correspond with the waterlines.

members forming the hull must be screwed to this backing piece after they have been screwed together, the various stages of erection being shown in the photographs.

General Procedure

The choice of timber is very important. A species with a close and silky grain that will carve or plane nicely in any direction without flaking or roughing is the ideal. No rules can be laid down upon a species, because samples vary so much, but white cedar, Douglas fir and beech are reliable common examples. Avoid resinous timbers like pitch pine, and although some hardwoods like American white oak, teak and satin walnut look suitable, an amateur will not be able to carve them exactly as he wishes, due to the marked and inconsistent grain. Mahogany and elm are generally useless. In all cases do not think of varnishing the hull on bare wood. There is no object in showing the waterlines, and with a painted hull, say white topsides and red bottom, the appearance will be more like a full-size hull and the lines can be more correctly criticized. With a painted finish the various members can be made from odd pieces of timber of different species which may be easier to find.

Certain American boat builders have used half-models to obtain their table of offsets, and in this case a very accurate reproduction of the lines is necessary, with all the waterlines showing. For this purpose they make the laminæ of alternate light- and dark-coloured woods.

Measure up the approximate sizes of all the pieces required, using a normal carpenter's rule on the drawings, and cut out rectangular pieces to these dimensions. The thicknesses must be exact, so next plane down each piece and test its thickness with calipers. Then clamp together all the members between the top and bottom-most waterlines, and caliper the overall thickness. It will probably

A HALF-MODEL

be slightly too great, so that a shaving must be taken off each lamina until the clamped thickness is correct.

The topmost piece, bearing the deck and bulwarks, requires smoothing to fit the top of the piece below it, and the bottom-most piece, which is usually an awkwardly shaped section, tapering from the lowest waterline down to the garboard rabbet, can only be made to fit the next piece above it. The waterlines may easily be transferred to the laminæ, either by placing each piece underneath the half-breadth drawing and pricking through the line or by means of carbon paper. Normally, the top waterline on each piece will be the wider one, so transfer only the top one, and then cut out to this line with a compass- or band-saw and trim up accurately with a plane or spoke-shave. Having shaped all the pieces to their respective top waterlines they should be assembled by placing in position on the profile plan so that the ends of the waterlines line up correctly. They should then be drilled and screwed together as indicated in Fig. 66 and a pencil run along the bottom of each to mark the true positions of the waterlines at these points. Dismantle again and plane down all the edges to a bevel between the top and bottom waterlines, which are now marked. Do not make this bevel flat for'ard, but leave a slight convex curve, the amount of which can readily be seen by looking at the body plan. Where this bevel is to be concave, such as in the after sections, it can be planed dead flat and hollowed later with a gouge, and similarly, in cutting out the big top piece, be careful of flare, which may make it wider in places than the waterline marked on its under-surface.

On reassembling the parts the hull will appear almost correct to shape, though somewhat wavy on the outside. But before commencing to smooth it off and carve the concavities aft, set to work on the bulwarks and deck.

FINISHING THE HULL SURFACE

This task could be simplified by adding the bulwarks separately afterwards, but if they are fairly high ones, with clean timber and sharp tools a much quicker job can be accomplished by making the bulwarks solid and chiselling down to the deck. Shape down to the top of the rail first by marking from the profile plan; then get the half-breadth shape of the rail, which will vary a lot from the shape of the waterline already marked on the bottom of the block, due to flare and tumblehome. It is a sound idea to cut down to the deck before shaping the sides of this block, as holding in the vice will be facilitated thereby. Unless a very precise model is required the deck camber may be ignored, as it would only amount to about $\frac{3}{16}$ in. of rise at the point of maximum beam. Mark the sheer line and then chisel a rebate between the sheer and the rail to emphasize the bulwarks, as shown in Fig. 66.

FINISHING THE HULL SURFACE

To get the final accurate shape of the hull prepare a set of templates, made from thin cardboard, as shown in Fig. 67, of the station lines on the body plan and mark their positions along the hull. By holding each template to its position on the hull the wood can be shaved down to the correct curvature between the waterlines. Remember

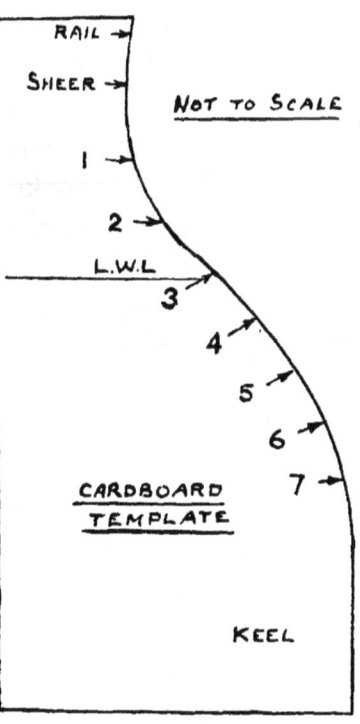

FIG. 67.—Check the hull with templates.

A HALF-MODEL

that the waterlines (or joints between the laminæ) are accurate and must therefore not be touched with the smoothing tools. Various tools can be used, depending on the curvature, but a plane, draw-knife and chisels are most handy on the for'ard sections, with a set of gouges for the after half. The large gouges should be internally ground. Having matched all the templates approximately, finishing can be completed with glass-paper.

As the extreme edges of all the laminæ are so thin, where the planking rabbet is represented, care must be taken in smoothing these places, but they are bound to flake and split slightly. These flaws can readily be hidden by stopping later, so the appearance of all thin edges may be ignored. Most other errors in carving can be remedied later, but bad ones may necessitate the renewal of a whole member.

Having marked the planking rabbet on the backing piece the bevels on the stem and sternpost may be applied, or the shaping of the cutwater and sternpost sides. The rudder is simple to make as a separate piece, but could be solid with the rest of the backing piece. When the hull has been screwed to the backing piece the appearance should be satisfying, and only painting remains. First fill the grain of the wood and any small irregularities in the surface by rubbing on a paste of plaster of Paris with a small wad of cloth. This will harden quickly and the surface can then be smoothed again with fine glass-paper. If flaws still appear, repeat the process.

An undercoat of pink priming or thin white lead paint all over is best, several coats being applied until an opaque finish hides all sign of the grain of the wood. Rub down all over with used No. 0 glass-paper after each coat. Then finish with a coat of white enamel or gloss paint above the waterline, and red or black below. Alternatively, flat paint could be used and finished with a coat

FINISHING THE HULL SURFACE

of copal varnish. A shade of ochre is best for the deck.

In describing the finishing processes it is presumed that the hull already appeared satisfactory to the critical eye, but even if by means of this half-model the set of lines used proved displeasing in reality, the model may be worth finishing for its ornamental value.

CHAPTER XIV
ELECTRICAL INSTALLATION

HINTS ON PROVISION AND MAINTENANCE FOR SMALL YACHTS

SOME sort of electricity supply is essential in almost any yacht, and a little extra trouble and expense will enable most of the appliances associated with electricity in the household to be adopted. The following notes are not intended as a complete guide, and any inexperienced reader who desires to carry out his own installation and wiring is advised to read a practical handbook on electrical wiring, or country-house lighting.

The whole wiring system must be carefully planned so that nothing need be added later, chiefly because the capacity (and number) of accumulators needed and the gauge of wire to carry the current must be calculated before the actual work is commenced. Batteries may be either of the nickel-iron (Nife) or lead-acid types, both having various advantages. The former type has a longer life and cannot be damaged by overloading. The latter type is cheaper, lighter in weight, and gives a higher voltage, so that fewer cells are needed to give the required pressure.

As an instance of an electrical installation we will take an extreme one, in which almost every possible piece of equipment is incorporated. The following shows the items which might be connected, with the approximate power taken in each case:

(a) Navigation lights and signalling lamp.   100 watts.
(b) Power winch, general purpose.   250 watts.
(c) Engine starter and ignition.   1,800 watts.

MAINTENANCE FOR SMALL YACHTS

(d) Wireless receiver. 60 watts.
(e) Motor for water pumps. 250 watts.
(f) Cabin lighting. 180 watts.
(g) Ventilating fans. 60 watts.
(h) Refrigerator. 150 watts.
(i) Plug points to take vacuum cleaner, immersion heater, cookery mixer, kettle, iron, etc. 2,000 watts.
(j) Clock, of negligible wattage.
(k) D.F. apparatus and radio transmitter. 200 watts.

This gives a total possible power output by the battery of 5,050 watts, but this load would never be imparted in its entirety; a maximum of 1,500 watts is nearer the calculation mark, as the engine starter may be ignored, all the plug point fittings would not be used at one time, and other items are used only intermittently.

Now taking the voltage of the supply as 50, and a current of 30 amps., the minimum capacity of the battery must be 150 amp.-hours. If a lower voltage were adopted, say 24 volts, the current would be 60 amps., necessitating very heavy wiring, and the cells would have to be of 300 amp.-hour capacity. Such big cells are heavy and expensive, so it would be even better to adopt 100 volts, using standard 70 amp.-hour car type batteries, all connected in series. The *same number* of similar cells would have to be used for the 50-volt supply, to make up *two* 50-volt batteries connected together in parallel in order to obtain a capacity twice that of each individual cell. With a 100-volt battery the wiring could be reduced in gauge, as the maximum current would then be only 15 amps.

Still taking an instance of the 50-volt supply, a $1\frac{1}{2}$ kilo-watt charging plant would be satisfactory, enabling the battery to be completely charged in about 8 hours. Any

ELECTRICAL INSTALLATION

faster charging rate might prove detrimental to the life of the cells. A 2½-h.p. engine would be needed, and would be quite economical, as the plant would normally need running only once every two or three days during the busiest periods.

Of course, this is rather an extreme case, and if only a few electric fittings were used, the main engine dynamo might be sufficient to provide all charging needs.

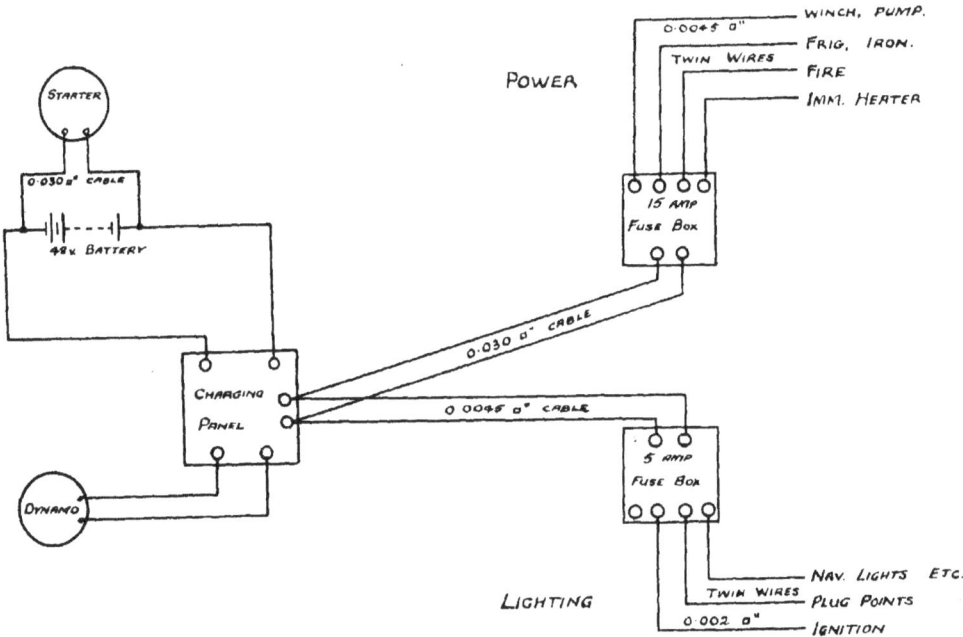

Fig. 68.—Skeleton diagram of electrical wiring.

Fig. 68 shows the complete wiring circuit, with the gauges of cable required for each separate circuit marked over the wires. The power side is divided into four circuits, each to carry 15 amperes, 7-strand vulcanized india-rubber cable, totalling 0·0045 sq. in., being used. The lighting side is divided into three circuits of 3-strand 0·002 sq. in. V.I.R. cable to carry 5 amps. This cable should be carried in screw-jointed galvanized-steel conduits. Alternatively, flat lead-sheathed twin cable could

TYPES OF INSTALLATIONS

be used, providing its protection is ensured. The starter motor must have separate leads to carry 36 amps. direct from the battery, and is in consequence of 0·030 sq. in. cross-section. The navigation light wires must be of flexible tough rubber covered cable, and should be provided with water-tight plug points at the deck.

By using 6-volt accumulators, a voltage of 48 will be produced by connecting eight of them in series, but this reduction of two volts will not prove detrimental to the working of the installation, and will mitigate too excessive a voltage when the charging set is in operation.

Types of Installations

There are three distinct types of battery installations suitable for a small craft:

(a) Two separate batteries with hand-operated change-over switches.
(b) Single battery with regulating switches, which enable a number of cells to be cut out both on the " charge " and " discharge " sides.
(c) Single battery without controls, as used in a car.

Type (a) is ideal, one battery being always ready for use and the other being charged or in reserve. The additional cost of a second battery is involved, but, if the extra weight can be carried, it will later be seen that the extra cost is well worth while.

Type (b) needs constant attention to regulate the voltage, particularly whilst charging. Additional cells are provided, so that the voltage may be maintained as the battery runs down. These extra cells, and others next to them, can be cut out as the voltage rises during the charge. The importance of maintaining the voltage at a constant level is not sufficiently great to justify this

ELECTRICAL INSTALLATION

complicated system, which will not be described in detail. If Type (*a*) is not adopted, Type (*c*) is the next best arrangement.

Type (*c*) is the simplest and cheapest system. No attention is needed, except to watch the reading of the voltmeter, which indicates when the battery is fully charged or discharged—in the latter case the dimming of lamps will be a warning. The disadvantage compared with the double battery system is that the charging should take place during idle periods when no load is being taken from the battery. If this cannot be arranged, there are certain points which must be watched so that damage to the electrical plant may be avoided. These will be described later.

Double Battery Installation

A typical diagram of Type (*a*) charging panel is shown in Fig. 69. This item would usually be obtained from the makers of the engine-dynamo unit and may differ in detail from the one described.

The double change-over switch is shown in the position to " charge No. 2 battery " and " discharge No. 1 battery." In the opposite position the No. 1 battery would be connected for charging and No. 2 connected to the load. The cut-out shown is of the type used in a car, which is actually a " cut-in " as well as a " cut-out." When the charging set is started, the voltmeter switch should be placed in the " dynamo " position, and as soon as the dynamo voltage rises to the correct value the contact " C " will close (operated by coil " P "). The cut-out will now remain closed by virtue of the charging current flowing through coil " S " and into the battery, and the voltmeter reading will be reduced according to the voltage of the battery. The ammeter (A1) will indicate the charging current.

DOUBLE BATTERY INSTALLATION

In the event of a failure of the engine, or if the speed drops so that the voltage of the dynamo is less than that of the battery being charged, the contact " C " will open. This arrangement is a safeguard and obviates the pos-

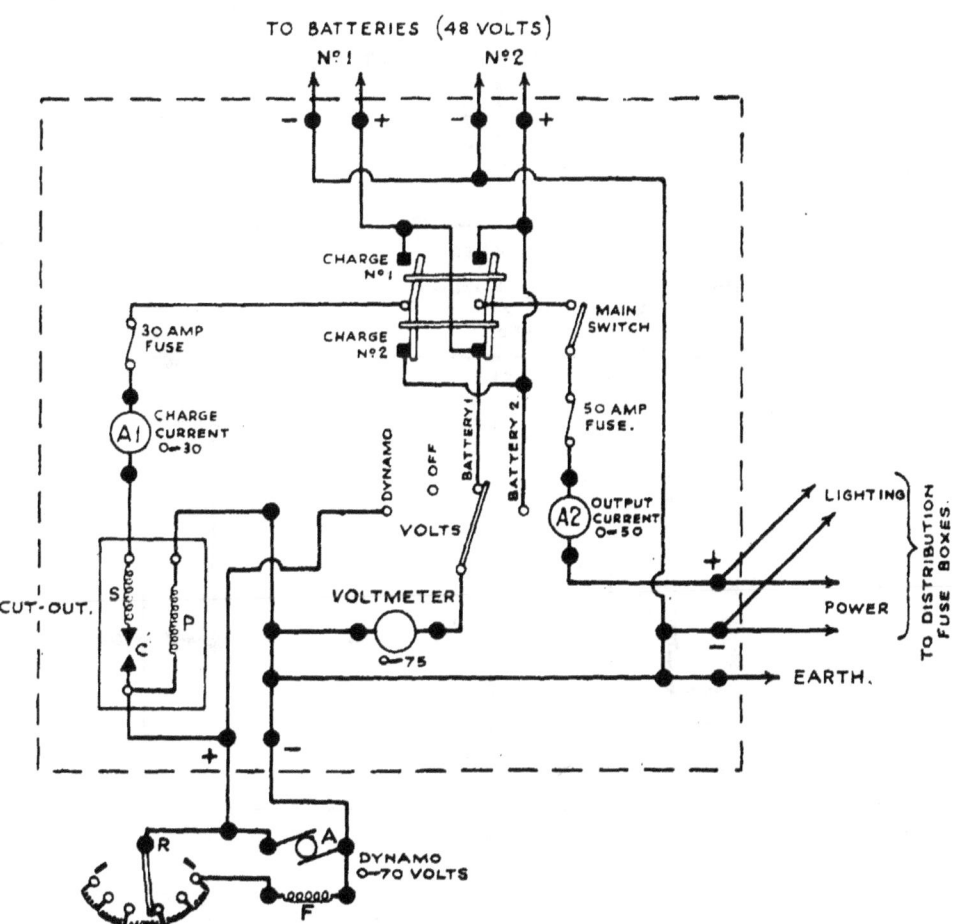

FIG. 69.—Charging panel—double battery.

sibility of a heavy current flowing back from the battery to the dynamo, which would then run as a motor. Should the cut-out fail to open, the 30-amp. fuse will blow. The panel fuses are usually of the cartridge type—keep a few spares.

ELECTRICAL INSTALLATION

Single Battery Installation

The simplest arrangement of the Type (c) charging panel is shown in Fig. 70. This differs little from that of a motor vehicle, with the addition of a voltmeter. The ammeter must be of the central zero type and should

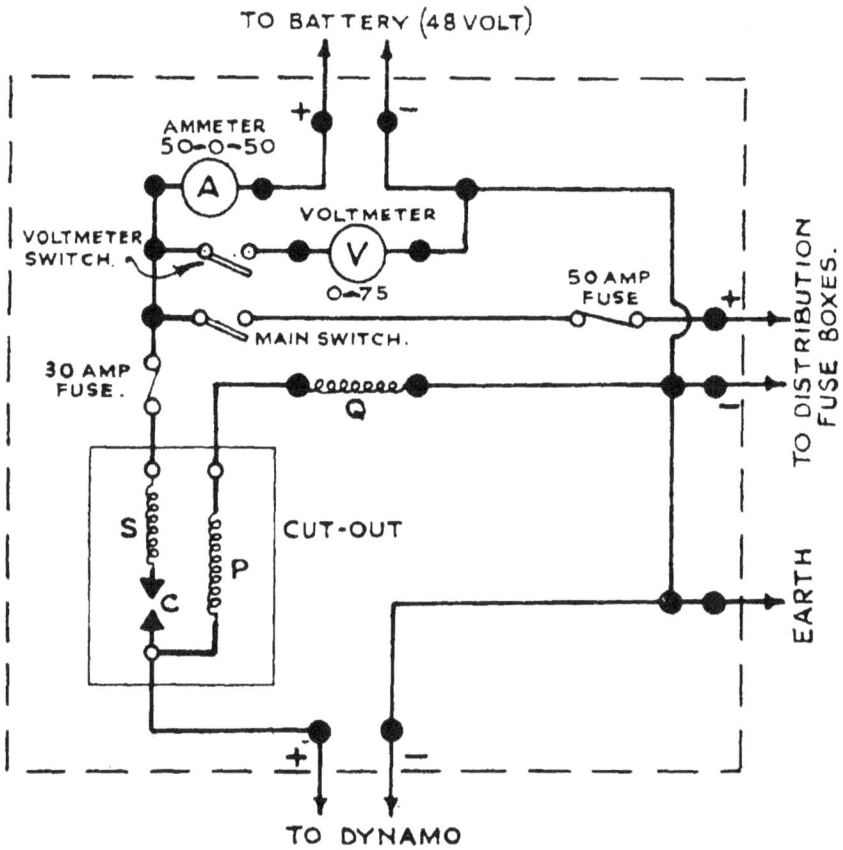

Fig. 70.—Charging panel—single battery 'A.'

read to 50 amperes. The cut-out is similar to that shown in Fig. 69. This must be wound for the correct voltage, to suit the battery, and the coil " S " must carry at least 30 amperes. If the reader should desire to make up this panel himself, a car " cut-out " may be used, but in this case the resistor " Q " shown in the diagram must be joined in series with the coil " P," the value of which (for

SINGLE BATTERY INSTALLATION

a 48-volt battery) will be three times that of " P " if the cut-out is of 12-volt size, and seven times " P " if a 6-volt cut-out is used. There may be difficulty in obtaining a car type ammeter of 50 ampere size, but a smaller one may be used, depending upon the maximum discharge current. A visit to the car breakers is indicated!

If difficulty is experienced in obtaining a " Charge-Discharge " type of ammeter of the desired size, the panel may be wired up as shown in Fig. 71, using a normal ammeter of 50 ampere size with a special 2-way switch. This switch must be of the " make before break " type, i.e. the moving arm must bridge the gap between the contact plates so that the charging circuit is not broken. When the switch is in position " D," the dynamo output current is shown, and in position " L " the load (current being taken by the lamps, motors, etc.). If the battery is being charged during an idle period, when no load is being taken, the reading in position " D " will indicate the charging current, but if any load is being taken whilst charging, the current going into the battery will be the difference between the two readings.

With the single battery type of installation no difficulty should be experienced if charging is carried out during idle periods. Whilst a battery is being charged the voltage rises gradually to (assuming 24 cells) about 55 volts during the first three-quarters of the charging period, and then, during the remainder of the charge—usually the last 2–3 hours—rises more rapidly to a possible maximum of 64 volts. This peak voltage is dependent upon the charging rate and can be minimized by reducing the output of the generator by means of the regulating resistance " R " (shown in Fig. 69)—this, of course, would lengthen the charging period: it is, however, beneficial to the cells.

ELECTRICAL INSTALLATION

If any electrical plant is in use there is no risk of any kind until the voltage rises above 55 volts. The following

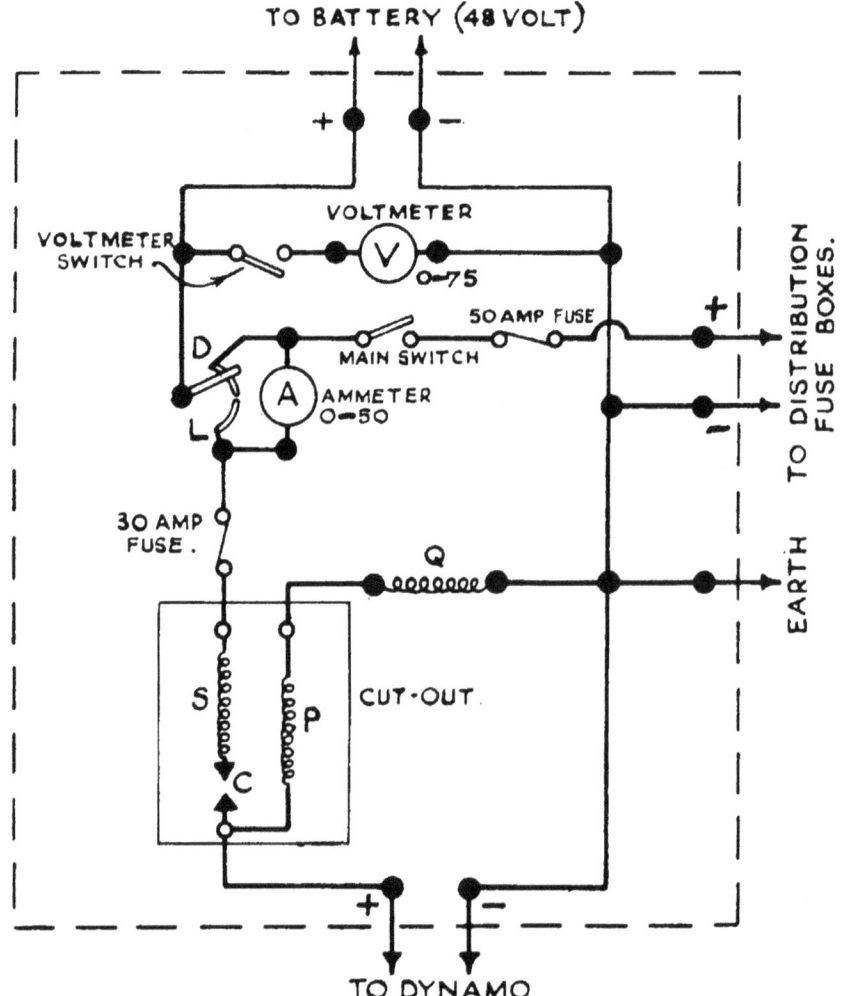

Fig. 71.—Charging panel—single battery 'B.'

notes cover points which must be watched when the voltage exceeds 55:

Lighting and Radio

The normal life of a lamp is 1,000 burning hours—this life will be reduced but, otherwise, the only effect will be a brighter light. The life of radio valves would also be reduced.

Power

Laundry Iron.—This will not be affected if of the thermostatically controlled type.

Refrigerator.—This, if the motor-driven pattern, is also controlled by a thermostat. The motor will run faster, but for shorter periods (see *Motors*).

Engine Starter, Ignition, Immersion Heater, Kettle.—These will not be harmed.

Vacuum Cleaner.—Do not use, except for very short periods.

Motors.—Any which are in use for more than short periods, such as ventilators, should be tested by placing one's hand on the machine. If the heat can be comfortably borne, the motor is safe.

It will be seen that, with a single battery installation, when the voltage of the battery on charge reaches 55, the best procedure is either to turn off the main switch or stop the generator and complete the charge at a more convenient time. Otherwise, attention must be given to the points enumerated.

EARTHING AND RADIO.

It is not considered necessary to adopt precautions against the possibility of electric shock on a 50-volt installation, but, when there is salt water about, quite an appreciable shock can be experienced under certain conditions with this voltage. To obviate this risk and to provide an " earth " for the radio, the following steps should be taken.

The metal-work of all electrical plant, such as the generator and all motors, the conduits or the lead sheaths of all cables, and the radio earth, should be connected

ELECTRICAL INSTALLATION

either to the " earth " terminal on the charging panel or to the negative main at any convenient point. In the case of portable items, such as a kettle or immersion heater, which usually have a " three-way " flexible cable, one wire of which is joined to the metal framework for earthing, a three-pin plug and socket should be provided, to prevent the possibility of reversing the plug, but the negative terminal should be strapped to the " earth " terminal in the socket.

Aboard small craft where wireless receiver or D.F. sets are to be used, and good reception will be required while a main or auxiliary engine with electric ignition is running, it may be found necessary to " screen " the set against interference from the spark plugs. In H.M. ships a special earth-plate is generally used. In the average yacht it will be sufficient to connect the negative main or the " earth " terminal of the charging panel to a keel bolt. The usual 7-strand bare copper earth wire will be suitable for this purpose.

Where interference with radio reception persists, despite precautions, the owner would be wise to take the advice of an electrical engineer, or, if a B.B.C. licence is held, the local Post Office Engineering Department, as the remedy may be entirely a matter of trial and error.

Wiring

It is not proposed to give full details of the electrical wiring, a skeleton diagram of which is shown in Fig. 68. The following notes, however, may be helpful :

Separate distribution fuse boxes should be provided for " power " and " lighting," of 15 ampere and 5 ampere size respectively. These should be of the " single pole " variety, i.e., only one fuse per circuit.

The cable is usually supplied in red and black colours. Red wire should be used for the positive main throughout

THE CARE OF BATTERIES

the system. This is the "live" main or "switch feed," and goes from the fuse in the distribution box to the switches in all cases. The black wire is joined to a common bar in the fuse box and goes to the lamps or one side of motors, etc. The "switch wire," which runs from each switch to the lamp or motor, is usually red, but black may be used for this.

THE CARE OF BATTERIES

There are certain very important points which must be watched in order to keep secondary cells in good condition, attention to which will prolong their life and ensure that their storage capacity is maintained.

A battery must not be charged at too high a rate—this must not exceed the "5 hour" rate, i.e. $\frac{\text{Capacity of battery}}{5}$. This rate, which would be 30 amperes in the instance of the 150 ampere-hour cells, would not be obtained from the $1\frac{1}{2}$ kilowatt generator suggested, which would probably only give about 24 amperes at the beginning of the charge.

It will be observed that, as the charge proceeds, the current—shown on the ammeter—gradually falls, and the voltage rises. When the battery is about 75 per cent. charged it is important to see that the charging current does not exceed the 9-hour rate, i.e. $\frac{150}{9}$, approximately 17 amperes in this case. If the ammeter reads 17 at this period the voltage should be about 55.

When the battery is fully charged, if still charging at the 9-hour rate, the voltage will rise to about 64, but this voltage will be less if the current is less. The main point is that when fully charged the voltmeter will remain steady, and this voltage and the reading of the ammeter

ELECTRICAL INSTALLATION

at the time should be carefully noted. As it is important not to overcharge the cells, the current should be adjusted, by means of the dynamo regulator, so that the ammeter reads this figure, and, when the voltmeter indicates one or two volts less than the noted reading the charge should be switched off. An overcharge once every three months is beneficial.

As soon as the fully charged battery is brought into use the voltage will rapidly fall to 48 volts, which will remain constant until nearing the end of the discharge, when it will fall fairly rapidly. It is harmful to discharge below 44 volts.

The batteries must be well supported on wooden racks so that they cannot move when the vessel heels. The acid must be " topped-up " periodically with distilled water. If any spilling of acid occurs, the usual level of half an inch above the plates may be reduced, but in this case " topping-up " must be done more frequently. All woodwork of racks and the floors under them should be painted with two coats of Solignum, applied hot.

The only real safeguard against the possibility of damage by sulphuric acid is either to stand each battery unit in a special deep lead tray or to line the floor of the battery compartment with sheet lead, which should extend to a height of at least four inches up the walls and a fixed board across the doorway. All lead-work should be beaten to shape in one piece, i.e. there should be no joints.

Before leaving a battery idle for any length of time, it should be fully charged. In any case, a " refresher " charge should be given every three months.

APPENDIX

TABLE 1

Copper Sheet for Sheathing

Imperial Standard Wire Gauge (S.W.G.) . .	22	24	26
Thickness in Inches .	0·028	0·022	0·018
Weight in Pounds per Square Foot . . .	1·292	1·015	0·831

APPENDIX

TABLE 2

COPPER COUNTERSUNK HEAD BOAT NAILS AND ROVES.

LENGTH	SIZES KEPT IN STOCK — WIRE GAUGE.													
5/8"	17	16	15											
3/4"	17	16	15	14										
7/8"	17	16	15	14	13									
1"	17	16	15	14	13	12								
1 1/4"	17	16	15	14	13	12								
1 1/2"		16	15	14	13	12	11	10						
1 3/4"				14	13	12	11	10						
2"			15	14	13	12	11	10	9					
2 1/4"				14	13	12	11	10	9					
2 1/2"			15	14	13	12	11	10	9	8				
3"					12	11	10	9	8					
3 1/2"							10	9	8	7	6			
4"								9	8	7	6			
5"									8	7	6	5		
6											6	5	4	
DIAMETER IN INCHES	.056	.064	.072	.080	.092	.104	.116	.128	.144	.160	.176	.192	.212	.232
ROVE SIZES	3/16"	1/4"	5/16"	3/8"	7/16"	1/2"	1/2"	1/2"	1/2"	9/16"	9/16"	5/8"	3/4"	

The dimensions and other details given in Tables Nos. 2, 3 and 4, are based on data supplied by Messrs. Guest, Keen & Nettlefolds, Ltd., Birmingham.

TABLE 3

Wood Screws

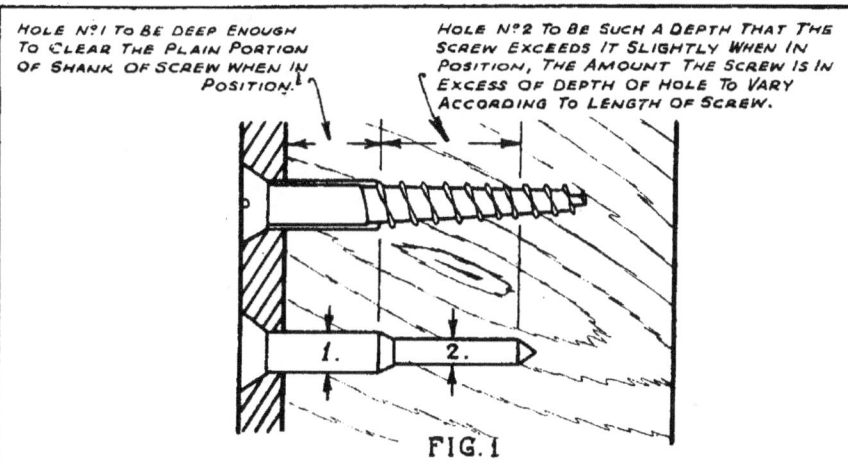

HOLE Nº1 TO BE DEEP ENOUGH TO CLEAR THE PLAIN PORTION OF SHANK OF SCREW WHEN IN POSITION.

HOLE Nº2 TO BE SUCH A DEPTH THAT THE SCREW EXCEEDS IT SLIGHTLY WHEN IN POSITION, THE AMOUNT THE SCREW IS IN EXCESS OF DEPTH OF HOLE TO VARY ACCORDING TO LENGTH OF SCREW.

FIG. 1

FOR METHOD SHOWN IN FIG 1 WHERE THE PLAIN PORTION OF SHANK OF SCREW ENTERS THE WOOD UNDERNEATH.

SIZES AND DRILLINGS.

SIZE OF SCREW S.G.	HARD WOOD ASH, ELM, OAK, BEECH, TEAK Etc.		SOFT WOOD FIR, PINE, SPRUCE MAHOGANY Etc.		SHANK DIAM: IN INCHES
	DIA OF BIT FOR DRILLING HOLE Nº1	DIA OF BIT FOR DRILLING HOLE Nº2	DIA OF BIT FOR DRILLING HOLE Nº1	DIA OF BIT FOR DRILLING HOLE Nº2	
4	1/8	HOLE Nº2 NOT REQUIRED	BRADAWL USED	HOLE Nº2 NOT REQUIRED	·108
5	1/8	"	"	"	·122
6	5/32	"	"	"	·136
7	5/32	"	"	"	·150
8	3/16	"	"	"	·164
10	7/32	"	"	"	·192
12	1/4	"	"	"	·220
14	1/4	5/32	1/4	"	·248
16	9/32	3/16	9/32	"	·276
18	5/16	3/16	5/16	"	·304
20	11/32	7/32	11/32	"	·332
24	7/16	1/4	7/16	"	·388
28	1/2	9/32	1/2	"	·444

TABLE 4

Wood Screws

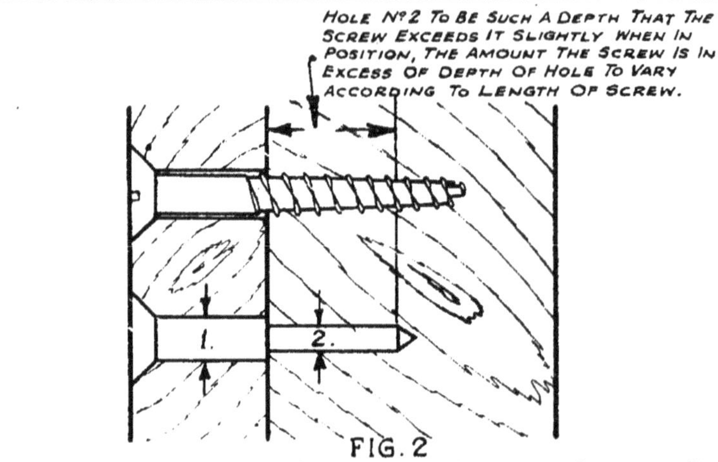

Hole No. 2 to be such a depth that the screw exceeds it slightly when in position, the amount the screw is in excess of depth of hole to vary according to length of screw.

FIG. 2

For method shown in Fig 2 where the plain portion of shank of screw does not enter the wood underneath.

SIZES AND DRILLINGS.

SIZE OF SCREW S.G	HARD WOOD ASH, ELM, OAK, BEECH, TEAK Etc.		SOFT WOOD FIR, PINE, SPRUCE MAHOGANY Etc.		SHANK DIAM; IN INCHES
	DIA OF BIT FOR DRILLING HOLE Nº1	DIA OF BIT FOR DRILLING HOLE Nº2	DIA OF BIT FOR DRILLING HOLE Nº1	DIA OF BIT FOR DRILLING HOLE Nº2	
4	1/8	BRADAWL USED	BRADAWL USED	BRADAWL USED	·108
5	1/8	"	"	"	·122
6	5/32	"	"	"	·136
7	5/32	"	"	"	·150
8	3/16	3/32	3/16	"	·164
10	7/32	1/8	7/32	"	·192
12	1/4	1/8	1/4	"	·220
14	1/4	5/32	1/4	5/32	·248
16	9/32	3/16	9/32	3/16	·276
18	5/16	3/16	5/16	3/16	·304
20	11/32	7/32	11/32	7/32	·332
24	7/16	1/4	7/16	1/4	·388
28	1/2	9/32	1/2	9/32	·444

www.ingramcontent.com/pod-product-compliance
Lightning Source LLC
Chambersburg PA
CBHW080538170426
43195CB00016B/2606